AF838169

FIRE BOMBERS IN ACTION

Videobook

FREDERICK A. JOHNSEN

specialty press
PUBLISHERS AND WHOLESALERS

Specialty Press
39966 Grand Avenue
North Branch, MN 55056
Phone: 651-277-1400 or 800-895-4585
Fax: 651-277-1203
www.specialtypress.com

Edit by Mike Machat
Layout by Monica Seiberlich

ISBN 978-1-58007-147-5
Item No. SP147

Library of Congress Cataloging-in-Publication Data

Johnsen, Frederick A.
Fire bombers in action / by Frederick A. Johnsen.
p. cm.
Includes bibliographical references.
ISBN 978-1-58007-147-5
1. Aeronautics in forest fire control. 2. Airtankers (Forest fire control)—United States. 3. Airtankers (Forest fire control)—Canada. I. Title.
SD421.43.J64 2010
634.9'618—dc22

2010000012

Printed in China
10 9 8 7 6 5 4 3 2 1

Front Cover:
Early test drop by the first Neptune air tanker, the Forest Service's N126Z, was made in 1969 while the patrol bomber still retained most of its original U.S. Navy markings. Local professional photographer Kenn Knackstedt documented a number of air tanker drops in the Medford, Oregon, vicinity starting in the 1960s, showing an artist's eye for composition and an enthusiast's enjoyment of the subject. (Kenn Knackstedt)

Back Cover, Top Left:
With its payload of 3,000 gallons of retardant being delivered through computer-controlled drop gates, this former U.S. Navy Lockheed P-3 Orion turboprop makes for a very potent air tanker today. (USFS)

Back Cover, Top Right:
Trading its usual garland of Douglas fir for Eucalyptus, Coulson Flying Tankers' classic Hawaii Mars rests on California's Lake Elsinore in October 2007. As of this writing, the former U.S. Navy flying boat still serves as a valuable fire-fighting asset nearly six decades after making its first flight. (Author photo)

Back Cover, Bottom:
Attesting to the progress made in aerial fire fighting, these photos show three generations of Douglas aircraft in action. A World War II-era B-18 Bolo performs a test drop (left), while a 1950's Conair DC-6B drops its retardant (lower right). Representing the Jet Age, a McDonnell Douglas DC-10 is shown unleashing its payload (upper right). (Left: USFS; top right: Darrell Brooks; bottom right: Frederick A. Johnsen)

Title Page:
This is it! The big show – if this book could use only one photograph to depict the golden age of aerial firefighting, this gritty image of Boeing B-17 Tanker 15 dropping retardant among tall timber and smoky slopes would be the definitive view. Let there be no doubt about the dangers involved, and the courage of the aircrews. (USFS photo by Jim Hughes via Donald Keller collection)

Facing Page:
The business end of a water-dropping Martin Mars flying boat. (Forest Industries Flying Tankers)

Distributed in the UK and Europe by
Crécy Publishing Ltd
1a Ringway Trading Estate
Shadowmoss Road
Manchester M22 5LH England
Tel: 44 161 499 0024
Fax : 44 161 499 0298
www.crecy.co.uk
enquiries@crecy.co.uk

CONTENTS

ACKNOWLEDGMENTS

My study of air tankers spans more than four decades during which many historians, photographers, air tanker pilots, and others interested in this exciting vocation have lent their help. With apologies for anyone inadvertently omitted, thanks go to Abbotsford International Air Show (and Ron Thornber), Air Tractor, American Aviation Historical Society, Jim Babcock, Peter M. Bowers, Darrell Brooks, Canadair, Keith Conrad, Coulson Flying Tankers, Rose Davis, Evergreen International Aviation, Forest Industries Flying Tankers, Malcolm Gougon, Laura Graham, Ian Grob, Don Hackett, Hawkins and Powers Aviation (and Dan Hawkins and Gene Powers), Carl M. Johnsen, Kenneth G. Johnsen, Sharon Lea Johnsen, Cecilia W. Johnson, Don Keller, Kenn Knackstedt collection (with thanks to Betty Knackstedt), Arnold Kolb, Keith Laird, Chandy Lemperle, Jim and Jean Morrow, Dennis Peltier, Christine Romero, Ken Shake, Penn Stohr, Ann U. Suter, U.S. Forest Service, Wenairco, and Asher Williams. And anyone who has ever spent time reviewing air tanker activities has undoubtedly enjoyed the efforts of William T. Larkins and Milo Peltzer, whose devotion to documenting air tanker history and operations goes back more than 40 years. Likewise, documentation by Scott A. Thompson helps explain and track air tankers. Dave Arnold and Mike Machat at Specialty Press enthusiastically encouraged and supported development of this project, making it possible.

ABOUT THE AUTHOR

Frederick A. Johnsen has been a student of air tankers since encountering a hulking orange-and-black TBM Avenger at the airport in 1961 when he was a child. The roar of converted warplanes earning their keep as firefighters was—and is—heady stuff for any aviation enthusiast. Fred pursued degrees in history and journalism from the University of Washington where he worked his way through college at the Kirsten Wind Tunnel. He has been a daily newspaper reporter, winning an award for wildfire reporting. Fred also served as editor of *Western Flyer*, consulting curator to the Museum of Flight in Seattle during its design phase, and founding curator of the McChord Air Museum at McChord Air Force Base, Washington. Since 1981 Fred has served as an Air Force historian and curator, NASA public affairs officer, and currently as director of the Air Force Flight Test Center Museum at Edwards Air Force Base, California. The author of more than 20 aviation books and monographs, Fred's articles and photos also have appeared in periodicals including *Air Force*, *Airman*, *AirVenture Today*, *Airpower*, *Aviation Week and Space Technology*, *EAA Magazine*, *Flight Journal*, *FlyPast*, *Koku-Fan*, *Society of Experimental Test Pilots Cockpit*, *Wings*, *The Seattle Times*, *The Seattle Post-Intelligencer*, *Associated Press* wire service, and others. His work is published concurrently online by some of these media. Fred's aviation video productions have been screened at major aerospace events and selected for showing to the President aboard Air Force One.

PREFACE

Air tankers, fire bombers, borate bombers. Everybody has watched, either on the evening news or perilously close in person, as a thundering aircraft barrels through smoky skies to discharge a huge bloom of pink-red fluid over a wildfire. The urgent drama has been part of the American landscape for more than a half century. The frequent inaccessibility of wildfire sites makes the use of aircraft especially attractive as a tool to subdue fires.

Evaluations with sprayers, small aircraft dropping small amounts of water, and even a B-29 releasing water-filled drop tanks and bomb casings suggested there was something to gain from fighting fires from the air, but it remained for serious tests of war surplus Stearman and N3N biplane trainers, TBM Avenger torpedo bombers, and then B-25s fitted with fluid tanks in the 1950s to make the case. Into the 1960s, the availability of World War II–vintage bombers fostered nomadic air tanker operators. In a time before the whereabouts of every old aircraft was known by aviation enthusiasts, the call for air tanker support could mean a fantastically surprising concentration of vintage metal that shifted airfields and reconvened throughout the summer months to serve wherever flames sprouted. A gradual winnowing of the World War II bombers due to age, crashes, and their increasing value as museum pieces coincided with a second surge of larger air tankers using postwar P-2s, P-3s, C-119s, and prop-driven Douglas four-engine airliners. Gone are the warbirds en masse, supplanted by postwar aircraft ranging from helicopters to jumbo jets. But the excitement and danger remain; along with an ever-increasing sense of urgency as more and more Americans move into rural areas each year.

The story of air tankers in firefighting applications involves science and technology; evolving techniques and equipment; skill and luck. Some observers argue the air tanker industry has been too long in the shadow of starvation budgets. Newer air tankers will require contracts lucrative enough to justify their upkeep. Some people yearn for easier access to federal firefighting contracts—a return to the good old days. This volume doesn't pretend to referee the ongoing debates. Decisions and direction may ultimately come from sources as diverse as the U.S. Forest Service (USFS), contract air tanker operators, and Congress.

The air tanker pilots I have come to know over the years exhibit resourcefulness, ingenuity, skill, and the gamut of attributes ranging from pride to humility. I've never met the air tanker pilot who considered himself a hero, or who wanted to be called one. And that says a lot.

Some technical notes: The users of firefighting aircraft sometimes make the moniker one word, as "airtanker"; others spell it "air tanker." Depending on the origin of documentation, it may be seen both ways in this volume. Multiple sources have been sifted to derive a list of air tankers by tanker call number and civil registration, and a roster of known air tanker companies in the United States since the early days of aerial firefighting. Like mining claims in the California gold rush, so many individuals and companies got into and out of the business that these lists may never be complete. Take them as a diligent effort to acknowledge the people and airplanes that have made up this industry for decades. The discipline of rotary-wing air attack, while bringing retardant to a fire, is different enough from fixed-wing air attack to warrant its own study. Therefore this volume concentrates on development and deployment of fixed-wing air tankers in the United States, with acknowledgment of significant air tanker technologies created in Canada.

While techniques and equipment for aerial firefighting have evolved over the last half-century, the debate over wildfire management has addressed the degree to which fires might be managed instead of snuffed, to allow nature to harvest easy underbrush fuel that could spare major forests later. No matter how frequently air tankers are called out, or stood down, in this or any other fire season, one tenet remains unchallenged since the 1950s: the best way to leverage the aerial advantage is to start tankers flying in the earliest stages of a fire.

With appreciation for all who make a career of wildfire management, this book is dedicated to the stalwart band of air tanker pilots, past and present, and those who support them.

Frederick A. Johnsen
July 2010

Chapter One

AIR TANKER HISTORY

The Ford 4-AT-E Trimotor N9612 (Tanker 2F) carried three 100-gallon tanks for firefighting when it was converted in 1957. It was outmoded as an air tanker but still useful as a utility aircraft when photographed at Missoula, Montana, in July 1965. *(Author Photo)*

It is difficult to imagine an invention more conducive to innovation than the airplane. To some observers, the very notion of assembling wood, fabric, or metal in a deliberate way to break the bonds of gravity by using combustible liquids to create the required energy for propulsion is audacious beyond belief. So it stands to reason the people who are attracted to flying machines are predisposed to being innovators in their own right.

Prewar Biplane Experiments

The peril of large-scale wildfires ravaging forests and prairies is not new. Before the western migration, perhaps as many as 12 million acres of wildland burned each season, touched off by lightning. As the western United States became increasingly populous in the early twentieth century, the Pacific Northwest was scarred by out-of-control fires that burned 3 million acres in Montana, Idaho, and Washington in August 1910. Spurred by high winds, the northwest fires of 1910 burned an area the size of Connecticut in two days. Seventy-eight firefighters and eight residents died in the fires of 1910. Smoke darkened the sky hundreds of miles away as 1910 became a benchmark year still visible in ancient charred stumps. Someone estimated the timber lost in the blaze of 1910 at 8 billion board feet.

"PUTTING THE FIRE FIEND TO FLIGHT"
Striking cartoon by Yardley in the San Francisco "Bulletin," indicating the successful use of aircraft in forest fire control, as demonstrated by the air patrol recently established in the State of California.

Demonized fire skulked away before the vigilant eyes of a spotter in a stylized de Havilland biplane in this evocative editorial cartoon by Ralph Yardley in the old* San Francisco Bulletin *newspaper shortly after World War I. (Reproduced in *American Forestry*, May 1921)

A fire lookout plots a position on a wildfire in this crafted composite image made in 1913 by using a photo of the Cabinet Mountains and a 10-gallon-hatted subject in the foreground, photographed in Washington, D.C. The western United States was still wild to many Americans at this time. (USFS/W. J. Lubken and E. S. Shipp; B. Adams did the superimposing)

To a growing nation only recently realizing its coast-to-coast dream, the loss of so much timber was a financial and strategic wound to be healed.

The fires of 1910 set in motion national wildfire policy for the first time. The policy idealistically expected all reported fires to be extinguished by midmorning the day following their discovery. Federal money was made available to preempt a repeat of the 1910 disaster. The result over many subsequent decades was the elimination of many wildfires that traditionally singed underbrush while leaving big timber standing. As more and more Americans entered the western wildlands, the fuel-rich forests were vulnerable to wildfires worsened by the availability of underbrush—the unintended result of decades of stopping every fire possible. If recent years have seen a trend toward managing some wildfires by letting them burn undergrowth, that policy was largely absent in 1954 when the American Forestry Association (AFA) reported the first time on record when the national forests in Oregon and Washington survived the fire season of 1953 with no individual fire exceeding 300 acres in size before being squelched.[1] That sounded like good news back then, with timber, wildlife, people, and property protected, albeit with the unintended consequence of continuing accumulations of undergrowth and downfall fuel that could later lead to fires of huge proportion and awesome destructive power.

The year after the 1910 fire disaster, the Western Forestry and Conservation Association (WFCA) entered contract negotiations with Curtiss for a trial aerial patrol, but the plan was abandoned before it could be consummated; the concept was probably too radical for the times.[2]

While foresters pondered how to check fires like 1910's Big Blow-up, California's magnificent Sierra Nevada Mountains provided the terrain to generate thunderstorms that sent incinerating lightning into the trees, burning resources and habitat each summer.

Fire crews then attacked the blazes from the ground. Their success depended largely on how quickly they could reach a nascent fire. Spotters on the ground and in the air could relay word of a newborn wildfire to firefighters. Air Service Curtiss JN-4 biplanes flew fire patrols over mountainous California beginning in June 1919, adding their wilderness reach to the effort. Larger de Havilland aircraft quickly replaced them. Limitations on radio effectiveness led the fire spotter crews to relay fire sightings by airdropped messages and even carrier pigeons. Budget shortages handicapped a promising patrol mission for the Air Service, with the de Havillands sometimes grounded for lack of gasoline.

Use of Air Service biplanes for forest patrols was not unique to California during this era. In its May 1919 issue, *American Forestry* reported: "Army airplanes and captive balloons will cover portions of the National Forests of California, Arizona, New Mexico, and other states this summer, to aid in detecting and suppressing forest fires. In compliance with an order from Secretary Baker directing the Air Service to cooperate with the Forest Service of the United States Department of Agriculture in this work, conferences are under way to determine where, and to what extent the air scouts will supplement the forest rangers." The forestry journal explained: "That there is a distinct and important place for aircraft in fire protection of timberlands is regarded by the forestry officials as beyond doubt, but experimental trial of methods and possibilities will have to be the first step. This is now being planned for the coming fire season." As early as 1919, the use of aircraft to attack fires directly was posited: "It is believed that bombs charged with suitable chemicals can be used with good results."[3]

Circa 1921, the use of airships to fight wildfires was contemplated by Los Angeles County. Vintage truck bed appears to be loaded with fire extinguishers and perhaps milk cans. (USFS Collection)

Less prescient, though undeniably imaginative, was an idea calling for testing "transporting fire-fighters by dirigibles from which ladders can be lowered to the ground." But the main thrust for aircraft in wildfire control in 1919 was to be fire detection. "At present the Forest Service relies for this partly on patrol, usually by men on horses, motorcycles, or railroad speeders, and partly on watchers stationed at lookout points. Aircraft have many points of obvious superiority for both classes of detection work," the forestry journal explained. "Airplanes would use wireless in reporting fires, as they have done in communicating with the artillery," the article noted. The time was only 16 years after the Wright brothers' first flight, and American experience in the air was both short and limited in breadth. "From the Army standpoint, the use of aircraft in protecting the National Forests affords a valuable opportunity for training flyers and developing further the possibilities of aircraft and the art of flying."[4]

De Havilland DH-4 biplane of the Army Air Service provided forest fire patrol capability from Felts Field on the east side of Spokane, Washington, circa 1925. (John E. Dean Collection)

A year later, in May 1920, the AFA reported: "U.S. Army aviators while patrolling their aerial beats along the Mexican border will take time to locate forest fires on the Coronado National Forest, in southern Arizona, under an agreement made by Major Ralph Royce, of the air service, and District Forester Frank C. W. Pooler of the Forest Service. Under this agreement, army aviators will be permitted to leave their border patrol in order to ascertain the exact location of forest fires, and in certain cases special patrol trips may be undertaken with the permission of the district commander of the air service... When forest fires are discovered by the airmen, notice will be telegraphed to the nearest forest ranger."[5]

Results from 1920 Air Service fire-spotting flights in California encouraged foresters. More than half of the 196 forest fires discovered and reported by fliers from the 9th Aero Squadron operating out of Mather Field, near Sacramento, California, were reported with coordinates that had an accuracy of one-half mile, guiding ground crews to the locations. Using radios the pilots expeditiously reported 83 fires. District forester Paul Redington was quoted in *American Forestry* magazine as saying the results of aircraft spotters in California during the 1920 fire season "demonstrates beyond doubt that Airplane Fire Patrol in California has been successful."[6]

That year, Air Service fire patrols began in California on 1 June, with three aircraft flying from Red Bluff, two aircraft flying from Mather Field, another two from Fresno, and two from March Field, near Riverside. These locations gave the fliers access to long stretches of the Sierra Nevada Mountains from northern to central California as well as other ranges to the south. In the north, where a wildfire blackened 25,000 acres in Lassen National Forest at Mill Creek, "a special reconnaissance plane, equipped with radio and with a forest officer for observer, hovered over the fire and actually directed the movements of bodies of fire fighters by wireless messages received right on the fire line." This aircraft also patrolled the completed, unattended fire line stretching 14 miles. "If reports from the air showed the line to be clear, the fire fighters were kept at work elsewhere; but if the observer wirelessed in that the fire had broken away, then a force of men was rushed to the spot and the fire corralled again."[7]

Again in 1921, aircraft described in one article as "de Havilland battle planes" searched for fires over California, the *American Forests* journal reported.[8] In the decade-plus since the fires of 1910 that caused creation of a national fire policy, fire lookouts populated remote forest regions. By 1924, Air Service patrols had been discontinued (at least in part as a funding issue) and, according to one account, because lookouts were on station all day, while aircraft could provide patrol coverage only a few hours at a time.[9] The summer of 1924 proved brutal to California forests, and gave rise to lyrics that were said to resonate with concerned citizens at public meetings. To the tune of "It Ain't Gonna Rain No More," new words described the state's plight:

A rancher had a pile of brush;
There was an awful breeze;
He touched it off to see it burn,
And killed ten thousand trees.

Oh, it ain't gonna rain no more, no more;
It ain't gonna rain no more;
The forest fires all burn like hell,
'Cause it ain't gonna rain no more.

The ability to leverage firefighters into remote locations was recognized in Canada during this period as well. The *Journal of Forestry* reported in 1922: "British Columbia will use seaplanes in its fire protection work this season. Experiments are to be tried of transporting expert fire fighters from one fire to another."[10]

John E. Dean, a pioneer member of the aviation industry in Spokane, Washington, remembered that in early 1925, the U.S. Forest Service (USFS) arranged with the Army Air Service to patrol northwest forests by air. Three de Havilland DH-4s were assigned this duty. Larger-than-life Spokane pilot Nick Mamer and two other Army reserve officers were hired to fly the patrols.[11]

That same year, the U.S. secretary of war suggested future fire patrols should be mounted by commercial fliers in arrangement with the Department of Agriculture in an effort to remove the government from competing with private industry. Nick Mamer took such a job.

National Park officer John R. White wrote an after-action report on a 1926 fire at Kaweah in California's beautiful Giant Forest in the Sierra Nevada Mountains. His efforts at scoping out the size of the fire included requesting the use of an airplane. Using a de Havilland dispatched to Fresno in the San Joaquin Valley west of the Sierras, White flew for two hours over fire-threatened areas, "getting a splendid view of the fire and an opportunity to study the terrain on the fire line which could only have been obtained otherwise by several days on foot and horseback," White reported. Back on the ground at Fresno, White "telephoned instructions for changes on the fire line and also laid off one hundred men whom I considered surplus. It is no exaggeration to say that this airplane trip saved the government several hundred dollars a day in labor alone for a period of several days."[12]

The staunchly pro-aviation *Aircraft Year Book*, a valuable resource for details on aircraft production and use for several decades, captured the essence of forest fire patrols for 1927. Flying over California, Oregon, Washington, and parts of Idaho and Montana, the USFS that year employed ex-U.S. Mail de Havilland 4 biplanes, flown and maintained by pilots and mechanics hired by the Forest Service. If the old DH-4s, descended from a 1917 British design, were long in the tooth by 1927, they were better than nothing. Army Air Service personnel inspected them monthly. Flying with a $50,000 budget, the Forest Service de Havillands logged 247 productive fire patrol sorties that year, amounting to 547 flying hours.[13]

The Ford 4-AT-E Trimotor N9612 (Tanker 2F) carried three 100-gallon tanks for firefighting when it was converted in 1957. It was outmoded as an air tanker but still useful as a utility aircraft when photographed at Missoula, Montana, in July 1965. (Author Photo)

Forest Service bird dog dips its wing to watch an early biplane air tanker—probably an N3N—release a small load over a wispy plume marking a fire in a remote mountainous area. (USFS)

A Stearman biplane takes on a load of pink-dyed bentonite. Pilot's helmet is adorned with classic World War II-vintage AN-6530 goggles in this early air tanker vignette. (USFS)

The USFS de Havillands flew on demand, not on schedule. They reconnoitered known large fires, confirmed reports of other fires in areas difficult to view by terrestrial means, and launched in the wake of thunderstorms. The authors of *Aircraft Year Book* wondered why the federal government did not further exploit the capabilities of aircraft in fire protection work. They speculated that limited budgets and staffing probably made it necessary for Department of Agriculture (USFS) to make either/or choices in fire management that did not embrace aviation as much as air-minded boosters might have wanted. Writing in 1928 about the year just past, they also expressed hope that aircraft would soon participate in experiments involving direct use in firefighting.[14]

Anecdotes describe Nick Mamer and a local Spokane Forest Service official testing the ability of water-filled beer kegs to withstand impact by kicking them from Mamer's Buhl aircraft over Spokane's Felts Field in 1930. The idea was to deliver water to ground firefighters. The duo also experimented with a hose trailing from a Ford Trimotor, spewing water over a test area. The volume of water was insufficient and ineffective, according to the story, but Mamer and the Forest Service official (probably Howard Flint) deserve acknowledgment for addressing the use of aircraft in fire suppression.[15]

A tongue-in-cheek account of the Spokane tests appeared in a Forest Service newsletter published in Missoula, Montana, in 1935:

> *Few persons in R-1 realize that, under the strong emotional urge created by the historic Selway fire, it was decided that henceforth we should take the matter of precipitation for forest fires into our own, self-acknowledged, capable hands. Why not carry water aloft in an airplane? It must come down. Only problems are to get it down in right time, right place and sufficient quantity. The undersigned was delegated to find out.*
>
> *First attempt was based on engineering plus psychology. If a well built, oak beer barrel would stand the shock? If smoke chasers heard a rumor that beer barrels were being plunked down from the air on Class-A fires? Get-away and travel time must be lessened. A brand new 8-gallon oak beer barrel filled with water; total weight 102 lbs. About 250 feet above the airport it was kicked out of Mamer's Buhl, forward speed about 75 miles per hour.* Poor keg! *We gathered it up from a quarter acre, no two staves left together. On the field a hole and a wet spot—looked as if someone had tried to drown out a badger.*
>
> *Next attempt, a specially built 18-gauge, galvanized iron can, brazed and reinforced at the ends, 8 gallons of water. Dropped from 250 feet, it broke like a fresh egg dropped on a concrete walk.*
>
> *A simple, 18-foot, Irvin cargo parachute let down two filled 8-gallon cans in succession—beautifully, gently, right side up, and close to the mark. A homemade chute of cotton sheeting with cotton fish cord suspension lines caused near-collapses among the audience. It opened, a line broke, and then 10 more broke in rapid succession.* Poor passenger! *Well, half the chute continued to function. The can punched a hole 6 inches deep in the field but was recovered still tight and undamaged. Chutes work. They are quite a lot of bother, cost something—$10 to $30—and will be torn by trees or snags.*
>
> *A specially built, welded, 16-gauge, black iron can holding 5 gallons of water stood the shock three times, bounces about 20 feet, is still good.*
>
> *Next trial: the Ford Trimotor—inside, a 100-gallon tank, lever-opening gate valve, 50 feet of rubber-lined cotton hose trailing out from a door opening, 30 lbs of sheet lead wrapped around the trailing end of the hose. The hose collapsed from the air blast so that little water went through it, didn't appear such a happy thought when it threatened to wallop the tail assembly of the plane. Women would have known better than to waste time with cotton hose. Then we got a reinforced rubber hose and put 5 lbs of white powder in the tank to make the water more conspicuous. Fine doings. A beautiful rain out of the end of the hose, 30 feet above ground—on the grass the lightest dew, covering a strip about 9 feet wide, half a mile long. Wouldn't have phased a light grass fire.*
>
> *One more trial: over green timber, a bigger hose, a nozzle to help concentrate the stream, careful flying down close to tree tops. A little of the fluid reached the forest floor; not enough to wet through a silk shirt. We knew beforehand that 100 gallons of water would cover 500 feet of fire line only about 0.05-inch deep.*
>
> *Forest Supervisors:*
>
> *How much water do you need per chain, or per 100 feet, of Class-A or Class-B fire line, if we sent it down kerplunk in 10-gallon containers and let your smoke chasers put it on the fire as they see fit? We could carry as much as 350 gallons of water or chemical in a Ford. We may yet drop some more water.*
>
> *—H.R.F.*[16]

If airplanes could fly over a fire, why couldn't they drop water on the fire?

Still other anecdotes credit California pilot C. J. "Red" Jensen with installing a pair of watertight hoppers on either side of the fuselage of a surplus World

War I biplane and making two drops on a fire east of Oroville, California, in 1931. Soviet tests in 1934 used an agricultural spray aircraft, and tended to confirm the ineffectiveness of spraying water or fire retardant. By 1937, the USFS was testing the release of 10-gallon tin cans filled with chemicals; some theoreticians of the day had concluded unrestrained bulk-water release would be spread so thin by the aircraft's slipstream that this method—later so successful—was not tested. Instead, the small tins were dropped. Impact caused a lead diaphragm in the tins to rupture promoting mixing of chemicals leading to discharge of foam fire extinguishing material. This was not effective.[17]

In wartime 1943, the Northern Rocky Mountain Forest and Range Experimental Station published an extensive paper by Roy Headley, titled "Re-thinking Forest Fire Control." Addressing the concept of aerial bombardment of wildfires, Headley described early efforts at using chemicals to control wildfires: "Experiments by the author in 1911 demonstrated that tetrachloride, a chemical used in a popular form of commercial fire extinguisher, had no measurably greater efficiency than plain water for forest fires. The exclusion of oxygen by a heavier-than-air gas seemed a 'natural'; but in effect it was found that a fire would have to be on level ground and surrounded by a wall if the gas were to be kept from flowing away before it could stop combustion. Even on a fire within a box, this gas had little superiority over water."[18]

Foresters did not overlook rapid advancements in the reliability and precision of aeronautics and chemical warfare during World War I. Headley recounted the post–World War I era, saying: "...after the first World War, it was natural that many men wondered why fires could not be 'bombed' with fire-retarding or fire-extinguishing chemicals. Col. William B. Greely, then Chief of the Forest Service, accordingly organized a cooperative study with the Chemical Warfare Service of the Army. The conclusion reached was that there was no worthwhile practicable use to be made of any known chemicals." But into the 1930s the USFS spent time and money researching ways to drop chemicals or explosives from aircraft to attack fires. Tests included the use of monoammonium phosphate as a retardant added to water. Results were not uniform, and indicated different concentrations had different effects on various types of fires. Foaming agents were tested as far back as the 1930s, but still the viability of air tankers remained an illusive goal.[19]

Forest Service testers in the 1930s also contemplated the effectiveness of explosive bombs, according to Headley, "which would throw dirt on the edge of a fire and thereby retard its spread." Though this pioneering work did not yield successful fire-attack aircraft at that time, it did contribute to proving the efficacy of dropping smokejumpers near a remote wildfire. Headley's paper cites the decisive obstacle to successful aerial firefighting at that time as "inability to hit the target." His 1943 study bluntly offers: "Whatever the

With its eight machine guns removed in peacetime, this military P-47N Thunderbolt (44-89326) carried two finned drop tanks when photographed during the 1947 Montana fire bombing tests. A U.S. Forest Service emblem adorns the engine cowling. (USFS via Craig Kaston)

truth may be about military 'precision' bombing, protracted effort failed to attain enough accuracy in bombing fires to make the method worthwhile."[20] This included chemical bombs; still the concept of free-fall liquids had not gained traction.

Headley allowed himself the freedom to be a futurist in his 1943 report. While saying no chemicals or aerial delivery methods had succeeded to date, he expressed hope that new rotary-winged aircraft would one day change the picture, allowing precise drops to be made. He imagined a list of firefighting agents that helicopters could release, including "chemical solutions or dusts, dirt-throwing explosives, diatomaceous earth, Portland cement, or even plain water."[21]

Postwar B-29, P-47 Drop Tank Tests

In the summer of 1947, a Republic P-47N Thunderbolt at Eglin Field performed a military test of firefighting theory by releasing a pair of water-filled 165-gallon drop tanks. In July, two P-47Ns and a B-29 flew to Great Falls Army Air Base, Montana, on the eastern edge of Montana's divide between open range and forest land, to expand on this approach to aerial firefighting. Test fires in the Lolo National Forest were targeted, followed by sorties against actual wildfires in western Montana. USFS emblems were applied to the B-29 and at least one of the Thunderbolts.[22]

The Thunderbolts dropped a total of 56 tanks; some fitted with stabilizing fins, as well as unfinned variants. The fighter-bomber pilots employed both glide-bombing and dive-bombing approaches to the fires. Finned tanks were used for the P-47 dive-bombing runs, which were discontinued in favor of glide-bombing passes with unfinned drop tanks. Tanks dropped by the P-47s were expected to rip open on impact, spewing water over the fire area. Initial results indicated the P-47 drops were superior to those made by the B-29. Funding problems caused the P-47Ns to stand down for a week in August for lack of gasoline. A proposal to deploy 75 fighters and 30 B-29s as fire bombers for the 1948 wildfire season went fiscally stillborn.[23]

The giant B-29 Superfortress, nicknamed The Rocky Mountain Ranger, cast a big shadow from its 141-foot wingspan over the Montana forests as it hefted eight 165-gallon drop tanks in its bomb bays. Weighing about 1,000 pounds each when filled with water, these tanks were armed with proximity fuses set to explode and rupture the tanks 50 feet above the ground to foster maximum dispersion of the water and some chemical fire retardants. The Superfortress dropped 46 tanks on test fires during seven missions flown typically at 3,000 feet. One tank could cover a swath 48 feet wide and 108 feet long.[24]

An Aerial Bombing Evaluation Board composed of seven forestry specialists and one lieutenant colonel from the Army Air Forces Air Proving Ground Command watched the tests. That summer of 1947, the euphoria over American success in World War II was aging on the vine. It was a transitional period at once filled with hope and uncertainty, and not yet infused with the Cold War urgency that would follow. Into this arena, a vastly downsized military could benefit from meaningful missions. The Aerial Bombing Evaluation Board opined that deployment of warplanes as firefighters "will help maintain high esprit de corps of military personnel in peacetime by assignment to productive missions that contribute to training."

Board members sent out a letter following the 1947 tests in which they confidently said: "After this brief survey we feel that this method of fire suppression offers definite promise for the better protection of lives and property." The Board bolstered its optimistic statement: "The present project already has proved that military aircraft can be flown in mountainous areas and that tanks containing extinguishing agents can be dropped with sufficient accuracy to hit and retard the spread of small fires."[25]

If hindsight questions the use of fused bombs over domestic forests, the Board already anticipated some issues. Their report referenced "sound and carefully followed publicity, which will guard against unfavorable public reaction, yet keep the field fully and currently informed."[26]

If the 1947 Air Force/Forest Service tests had limited funding, and if results with drop tanks were not optimal, the notion that aircraft could fight fires was nonetheless reinforced over the forests of Montana.

Though the concept of aerial firefighting remained tantalizing, it was not yet deemed practical. Water bombs weren't the answer, and according to one USFS official, the value of wildland resources had not yet appreciated enough to overtake the considerable costs of retardants and aircraft to deliver them. That began to change in the 1950s.[27]

The use of a four-engine B-29 in the Montana experiments pointed the way to the value of larger aircraft. Six years later, modern air tanker technology was suggested by serendipity in the latter half of 1953. The prototype Douglas DC-7 four-engine airliner was fitted with water ballast tanks to enable flight-test crews to alter the aircraft's center of gravity for handling-qualities testing. On one test flight, the Douglas flight crew jettisoned 1,300 gallons of free-falling water over the runway at Palm Springs, California, laying down a wet swath 200 feet wide and nearly 1 mile long. The implications were obvious, and exciting. In early December 1953, with an isolated demonstration fire burning safely on the clay bed of Rosamond Dry Lake in Los Angeles County, the prototype DC-7 roared over at altitudes ranging from 150 to 600 feet, dumping about 300 to 500 gallons on each pass.

A contemporary report noted: "This experiment showed that free-falling water released in sufficient quantity from low-flying aircraft will reach the ground in appreciable amount." Rain gauges located along the drop path recorded moisture accumulations. Length of the drop pattern was about 4,000 feet, with width varying between 125 and 300 feet depending on drop altitude. Airspeed was 150 mph.[28]

In early 1954, Marine Corps Reserve pilot Maj. Warren Schroeder tested a different free-fall drop system while flying a Douglas AD-2 Skyraider. The Skyraider carried a 250-gallon napalm tank, the ends of which were modified to use glass plates that could be broken by detonators activated by the pilot. A solution of water and foam was released in this way from a height of about 35 feet at 160 mph. The Skyraider placed a visible pattern on the ground 50 feet wide and 300 feet long, with the highest concentration of foam reaching 1/2- to 3/4-inch deep in a 15-foot-wide strip near the center of the pattern.[29]

A surplus U.S. Navy TBM-3 Avenger torpedo bomber, still bearing insignia and markings, begins to shed its former identity and military gear as it metamorphoses into an early air tanker in Arizona. In the distance, beneath the tip of the right horizontal stabilizer, can be seen other pieces of warplane salvage including parts of an Erco bow turret from a PB4Y-2 Privateer, removed when that bomber became an air tanker. This was many years before such parts had value in the restorations of many of these same aircraft. (Ken Shake)

Trio of bulky TBM Avengers belonging to Aerial Applicators of Salt Lake City in September 1971 typify the most successful first-generation single-engine air tanker. Available in quantity in both the United States and Canada, TBMs were prominent in the successful development of air tanker operations. (Author Photo)

Reeder Flying Service operated Tanker 23, a bright yellow TBM Avenger photographed in mid-drop. (USFS)

1950s Avenger Experiments Lead the Way

In 1954 federal, state, and local agencies teamed with the University of California and contractors, including Paul Mantz Air Services, to explore all manner of equipment that could be brought to bear on the seasonal wildfire problem. Mantz, famous as a motion picture pilot and sometimes aerial chauffeur for Hollywood stars, used a TBM-1 Avenger modified by lining the torpedo bay with a two-compartment plywood tank. Plywood bomb bay doors closed beneath the two tanks. The use of plywood was an expedient to enable quick prototyping of the system. The rigged system was made watertight by installing a large weather balloon in each bay—a test expedient. Toggle switches in the cockpit released bomb shackle mechanisms that caused the plywood doors to fall open in flight, quickly rupturing the thin weather-balloon membrane. Total capacity of the tank was 600 gallons—verging toward a respectable load size for air-attack fire suppression.

Drop tests ranged in altitude from 10 feet to 100 feet, with the best fire suppression pattern realized at 50

Tanker E25, a TBM from Sis-Q Flying Service, pauses in a puddle of retardant at Ukiah, California, on 20 August 1967 during repeat sorties against a nearby wildfire. This TBM later crashed in Quebec in 1975. (Kenneth G. Johnsen)

On a 1965 July afternoon in Montana, these five TBM air tankers sit on the line at Missoula. Several canopy and windscreen combinations are in evidence; two TBMs closest to the camera have wraparound windscreens to remove obstacles to the pilot's field of vision. (Author Photo)

feet. At this height, the Avenger could cover an area about 90 feet wide and 270 feet long. On 1 September 1954 the experimental TBM made two drops on an operational fire—the first free-falling, large-volume water drop on an actual blaze in the United States.[30]

The USFS obtained its own surplus TBM-3 Avengers in 1956 and began converting them in 1957. A 440-gallon tank constructed at state shops at Davis, California, led to a larger 600-gallon configuration. The Avenger's hydraulic system provided normal actuation for the drop doors; in an emergency, two 3-inch (later 6-inch) drop doors could be opened by the pilot tugging on manual release handles.[31]

The use of air tankers was gaining momentum throughout the west by 1957. The underlying premise—initial attack while a blaze is small—proved the worth of speedy air tankers vaulting over impassable terrain. But the tankers also supported ground fighters as they worked large fires. In one instance, 800 gallons of borate slurry and so-called wet water strategically dropped in the path of a crown fire turned the tide. A crown fire blazes in the tops of trees and readily leaps from tree to tree as hot embers are blown on the wind. Difficult for ground fighters to quell, this particular crown fire was forced to the ground when it hit the borate-treated area, enabling a team of 75 ground firefighters to gain control.[32]

By 1958, air tankers doubled the volume of fire retardant they had dropped the previous year, tallying the new record to about 1.441 million gallons.[33]

In 1961, near the site of the 1953 DC-7 experiments, the U.S. Army aviation test facility located on Edwards Air Force Base conducted an evaluation of a typical TBM Avenger torpedo-bomber/air-tanker conversion. Using a serviceable 600-gallon tank, Avengers formed an important component of the early air tanker fleet, and the Forest Service sought to quantify the TBM's capabilities and shortcomings. (TBM test results are summarized in the TBM section of Chapter Two on page 29.)[34]

By the late 1950s, the model for aerial fire suppression was well formed by tests and actual operations with aircraft like the ubiquitous TBM. Surplus sales of larger bombers launched a nomadic and varied industry in America. Their built-in bomb bays provided convenient space for fluid tanks near the crucial center of gravity. Still other surplus aircraft, lacking internal bomb bays, would be modified with purpose-built internal and external tanks to get the job done.

Better Than Water

Early California air tankers in 1955 and 1956 dropped water, which could dissipate on the way down over a hot fire. An emulsifier could improve drop coverage; early operators recall trying mixes of jello before arriving at borate.

Scientists and foresters speculated at least as far back as 1919 that chemicals would provide better fire suppression than could be obtained by dropping untreated water on wildfires. The quest for firefighting chemicals took particular requirements into consideration: Retardants had to be effective, available, transportable, and affordable. Pioneering the use of chemicals, United States Borax and Chemical Corporation, the 20-mule-team company, developed Firebrake by 1956, using sodium calcium borate as the fire retardant. Though subsequently replaced by other chemical compositions, borate lent its name alliteratively to air tankers, known for many years thereafter as "borate bombers." A 1960 advertisement for Firebrake described the product: "It has been used to knock down and hold small fires and, in other cases, to establish chemical firebreaks for the close support of ground crews. So you see, whether Firebrake is cascaded from the air, or is utilized for ground attack, it's effective...and it buys time! Firebrake... is quickly mixed in the field with water, as a slurry, and remains stable. White in color, it is easily discernible after application because it clings to whatever it touches as a protective coating which stubbornly resists flame."[35]

Scientifically, sodium calcium borate powder is only slightly water soluble. When mixed it forms a thixotropic slurry. This is a watery mixture of undissolved material that, when motionless, becomes thick and stable, resisting separation into liquid and solid components. When stirred, it becomes more fluid.[36] The pinkish-white borate powder had abrasive properties that could damage pumps; retardant base operators benefited from using harder metals in pumps than brass.[37]

The white trace left by borate gave rise to a new club: Firefighters and others who were hit by falling borate earned membership in the Seagull Club. Firefighters were told to avoid being hit by falling slurry, which can knock down small trees, break limbs, and push debris ahead of its path. Early advice was to drop flat on the ground, behind cover if possible, if in the path of a retardant drop. In at least one instance, however, an early-day air tanker pilot deliberately skewed his drop to drench a fire crew when he observed a treacherous shift in the fire's path that cut off the crew's only escape. After the fire swept around them, the firefighters sustained only minor burns and were able to escape the lethal firetrap.[38]

A report printed in 1960 by the California Division of Forestry gave borate its due: "Sodium calcium borate, a refinement of one of the many chemicals tested at Operation Firestop, was the first retardant to be put into general operational use. Borate proved to be an effective fire retardant. It is the material with which the air attack program was pioneered."[39]

Even as borate was proving itself in the field, California and USFS experimenters evaluated the efficacy of an alternative thixotropic slurry of water mixed with swelling bentonite clay. Bentonite clay, used in oil-well drilling operations, made a stable gel in water. Bentonite retained water for two or three hours after the slurry had been released over vegetation. Motivation for using bentonite included price, estimated as one-tenth to one-twelfth the cost of borate. Bentonite could be a logistics boon, as it required less shipping and warehousing because its dilution ratio with water is only one-fifth that of borate. Bentonite slurry gave essentially the same drop pattern characteristics as did borate. Tests conducted in 1958 on test burns, as well as real-world air tanker application on wildfires, indicated bentonite to be a good fire retardant for two to three hours after application; but borate had greater longevity after being applied to vegetation. Lacking borate's stark white color, bentonite blended in with its surroundings, and air tanker pilots found it difficult to see where previous drops had been made. Out of this experience came the use of red-pink aniline dye for fire retardant. Experimentation with yellow dye showed it to be visible against green vegetation, but useless on dry grass, and efforts at dyeing bentonite white were not feasible.[40]

Visitors to Missoula, Montana, inspect a deep-bellied TBM air tanker and what appears to be a livestock watering tank set up as a retardant ground container in the early years of fire operations. Tanker operators have long shown individuality in painting their aircraft a wide array of bright colors to make them more visible over the congested, smoky scene of a fire. (USFS)

The 1958 bentonite tests on actual fires included its use in northern California of a diminutive 120-gallon Stearman biplane tanker and a slightly more capacious 180-gallon N3N biplane. TBM Avengers tested bentonite versus borate on Southern California's Monrovia fire that year.[41]

TBM Tanker D1, fitted with wingtip endplates, kept Visual Flight Rules (VFR) distances as it cruised past a large cloud during fire operations in 1967. (USFS)

Air tankers flew comparative test drops at Willows, California, in 1956, Santa Ana, California, in 1957, and Ramona, California, in 1959. Tests showed increased speed and altitude diminished retardant delivery accuracy and effectiveness; the higher and faster the drop, the more the retardant tended to break up into large drifting clouds instead of meaningful, wet ground coverage. A rule of thumb for effective retardant delivery has been described as falling like rain, with negligible forward motion. The development of various-sized air tankers was not simply a function of availability of surplus aircraft. Testers quantified the effectiveness of everything from N3N biplanes to PBY Catalinas and F7F Tigercats

BEST OF THE BEST, 1959

A Forest Service study of air tanker aircraft available in 1959 ranked nine types in descending order of desirability. Factors included minimum field requirements, airspeed, retardant capacity, slow-speed performance, maneuverability, pilot visibility for dropping and maneuvering, field maintenance, degree of pilot skill required, ease and cost of tanker conversion, and availability of aircraft. Listed in order of desirability, nine air tankers were ranked, with the highest marks going to the ubiquitous TBM, followed in order by: PBY, F7F, PV-2, PB4Y-2, B-25, B-17, B-26 (A-26), and YC-97.[64]

The industry was still in its youth in 1959, and tanker desirability changed over the years. The most favored single-engine TBM was phased out for multi-engine aircraft in the 1970s in the United States, while the next-to-last ranked A-26 remained in service well into the 1980s. (Appendix E reproduces the 1959 matrix that rated these and other available air tanker types.)

During the 1959 California study, lead-airplane pilots were asked to estimate what percentage of the drops made by three key air tanker types was effective (accurate) against an average target. The study reported TBMs had a 90-percent accuracy; PBYs 80-percent accuracy; and B-25s 60-percent accuracy. Though admittedly subjective, these figures offer a glimpse into the world of early air tanker operations as seen by the practitioners of the time. Higher accuracy translated into lower delivery costs since less retardant was wasted on bad drops.[65]

Consolidated PB4Y-2

Grumman TBM

Consolidated PBY

North American B-25

Boeing B-17

Lockheed PV-2

Grumman F7F

Douglas B-26

and reported in early 1960 that a large-capacity tanker flying low and slow is more effective than a small tanker. However, when terrain or stalling speed force the large tanker to drop higher and faster, it may not do as well as a small and more maneuverable air tanker.[42]

As bentonite moved into operational use alongside borate by 1959, researchers pondered another issue: In concentrated doses, borate temporarily sterilized the soil. This could inhibit grass growth for up to three years. Following a fire, it is desirable to establish new vegetation quickly to thwart erosion of the ground denuded by the fire. Phosphate-based liquid fertilizers showed promise as both fire retardant and subsequent vegetation enhancer. Eventually borate was sidelined for phosphate-based retardants.

Also tested by 1960 was algin, a water thickener composed of sodium alginate derived from Pacific Ocean kelp.[43] Algin-thickened water builds up a thick layer, which absorbs substantially more heat energy than does plain water. Once the water has evaporated following an algin/water application, the remaining algin dries to form a barrier film that helps seal off the

There's something leisurely about this cross-country formation with TBM Tanker A12 and PB4Y-2 Tanker B22. The TBM pilot has his bomber's canopy cranked open. Privateer Tanker B22 (later Tanker 122) would not outlive its firefighting career; it was damaged and written off in California in 1980. (USFS)

fuel from oxygen, retarding or preventing further combustion and rekindling of the fire. A TBM test drop of 600 gallons of medium-viscosity algin gel from a height of 125 feet at a speed of 120 knots gave a remarkably controlled pattern 350 feet long and 40 feet wide, with minimal drift, attributed to the viscosity.[44] Subsequently, algin may have found greater use with ground tank trucks and helicopters.

The efficacy of retardants dropped from the air accelerated the use of air tankers rapidly. A USFS division director reported that in 1956, retardant dropped on USFS fire operations amounted to only 23,000 gallons; by 1961, USFS air tanker retardant use exceeded 7.5 million gallons.[45]

The Monsanto Company marketed Phos-Chek, a diammonium phosphate retardant mix, in the 1970s. The company described Phos-Chek in advertising: "...Phos-Chek fights fire by absorbing heat and depriving fire of its fuel... The chemical's bright red color lets ground crews and pilots quickly locate areas that have been treated. And, when the fire's over, Phos-Chek helps fertilize the ground to promote new plant growth."[46]

A TBM Avenger pulls up and away from its retardant load in the rugged American West in this 1967 photo. (USFS)

A long-term retardant, such as Fire-Trol, uses a chemical salt like ammonium phosphate or ammonium sulfate, which alters the fire's combustion characteristics. The salts foster the formation of water, carbon monoxide, and char at a lower temperature than that of an unaltered fire, creating fewer flammable gases. Other constituents of long-term retardants such as Fire-Trol include clay, guar gum, and water, which cool the fire and the fuels. Phosphate is traditionally more expensive than sulfate, but it is also effective in smaller amounts per gallon of retardant mix. Therefore, it may be viable to use a higher concentration of cheaper sulfate to do the job.[47]

The salts that make long-term fire retardants effective can also be unwanted sources of corrosion. Retardants such as those made by Fire-Trol Holdings include corrosion inhibitors to keep the mix within USFS standards for corrosiveness. Fire-Trol can be supplied as a liquid concentrate or a powdered mix.[48]

Another tool for firefighters is Fire-Trol Firefoam, which reduces water surface tension to enable water to penetrate into some wildland fuels. Firefoam-treated water coats the fuel source with a layer of foam that absorbs more heat than plain water can absorb.[49]

Seaplane scooper air tankers sometimes carry a gelling agent aboard that can be mixed with the water they have scooped. Gelgard, a Dow Chemical product, is a powder polymer that can, when mixed with water, yield viscous water or gel depending on the quantity of powder applied. The powder absorbs water, but does not dissolve in it. The firefighting benefit of Gelgard is its ability to make water cling to surfaces as a viscous mixture. Once the water evaporates, Gelgard has no other retardant properties.[50]

Retardants in use today can have three basic effects on fuels to slow the fire: The thick consistency can coat the fuel, thereby robbing the fire of fuel; chemicals in the retardant can modify the nature of the fuel, inhibiting combustion or decreasing combustion intensity; and retardant can promote fuel cooling where the ambient air temperature is reduced by the evaporation of the water, also reducing the temperature of fuel, making it more difficult to ignite. At some tanker bases, the availability of long-term and short-term retardant gives planners more choices. Short-term retardant ceases being effective when its water content evaporates, typically one or two hours after application; long-term retardant is necessary if follow-up ground forces are more than an hour distant from the airdrop. But short-term retardant is substantially more economical than long-term retardant, and planners are urged to match the correct retardant with the situation at hand.[51]

Containerized High-Altitude Drop Test

In 1971, the USFS and The Aerospace Corporation used a PBY Catalina air tanker over Ryan Field at Hemet, California, to try drops of fluid in small packets from a height of 750 feet. As many as 5,000 water-filled, polyethylene packets tumbled from the Catalina. Drops from heights above 500 feet can be advantageous from an aircraft safety standpoint, but uncontained bulk liquids disperse and evaporate too much to be effective from such heights. Hence the use of small plastic bags to contain retardant during free-fall. The bags were meant to rupture on impact.[52]

Calculations indicated that polyethylene bags 1.5mm thick and measuring about 3 inches by 5 inches, filled with 4 ounces of liquid, could perform well in flight, and remain safe on impact by not building up too much energy. For the test, water dyed blue was inserted into the bags, which were then heat-sealed. The 5,000 bags were transported in drums filled with a water slurry to cushion them during transportation to the airfield. Hemet Valley Flying Service provided the PBY for the test (a TBM was considered and rejected because of difficulty in loading the bags into its retardant tanks). The design of the PBY's tanks in the patrol plane's fuselage permitted easier access for this unorthodox loading.[53]

The PBY was to drop at a speed of 120 mph at an altitude of 750 feet above ground level. Foam pads and kraft paper provided visualization of the drop and splash pattern. Testers observed the bags were affected in their fall by entrapment in the cushioning water slurry that protected them in the PBY's tanks. The slurry exited with the bags when the drop doors were opened. The test bags' rectangular shape also foiled some measurement predictions as some bags went broadside into the line of flight, presenting increased drag and less forward motion. Testers acknowledged more study was required on the storage and handling of large quantities of liquid-filled bags. Bags with better aerodynamic stability, and made of biodegradable plastic, were desirable.[54]

Although occasionally revisited, the notion of containerized retardant drops remains out of the mainstream.

Safety on the Ground

State and national forestry officials in California used strain gauges attached to horizontal and vertical plates to characterize impacts from retardant drops during 1957 tests at Orange County Airport in Southern California. They reported: "Impact forces from drops

A tank truck for Monsanto Phos-Chek retardant helped air tanker mobility during the fire season. (USFS)

Brightly dyed Phos-Chek retardant spews into a collapsible holding tank for pumping into tankers. (USFS)

A PV-2 Harpoon makes a low-level drop, possibly as part of the knock-down tests the USFS conducted to determine safe release heights for different aircraft, to avoid injuring firefighters on the ground. (USFS)

below 25 feet can be considerable, particularly in the horizontal direction," producing instantaneous acceleration of about 4 gs.[55]

The impact of fire retardant released from an aircraft can knock a firefighter down, or push dangerous debris ahead of its path. In June 1960, the Forest Service's Arcadia Equipment Development Center stood an instrumented crash test dummy in the path of several air tankers in an effort to quantify the hazards faced by ground firefighters from retardant drops. Air tankers sampled for the tests were an N3N, TBM, F7F, SNB/C-45, and PV-2, providing variations in volume and tank design. The Arcadia center explained the procedure: "Direct water drops were made on an instrumented dummy. Our dummy, 'Sierra Sam,' was equivalent in stature and build to a husky man, and was equipped with accelerometers inside his chest." Sam was placed in a standing position to receive drops at ever-lower altitudes until significant damage became apparent. Next, drops were made on Sam while he was in a prone position. All five of the air tankers performed these tests.[56]

It quickly became apparent that the impact on a standing firefighter included serious risks from the subsequent fall to the ground. "When knocked from the standing position, high values of 'g' loading were obtained in some cases just from falling down," the report noted. When the dummy was lying down, testers found "he was carried with the liquid on very low drops and thrown to the ground." This abrupt end to the dummy's bodysurfing adventure could lead to injuries. But the prone position proved vastly superior protection to the dummy for drops made above 15 feet in altitude. "Below this height a veritable wall of water carries the body easily. A hardhat on the dummy, with the chin strap fastened, usually remained in place and afforded good head protection."[57] Fortunately, airdrop tactics typically developed optimum drop heights greater than 15 feet.

With a pale yellow vertical fin, Tanker 71 looked like this at Medford, Oregon, circa 1967 to 1969. (Donald Keller Collection)

Without the yellow vertical fin that characterized it for a time in the 1960s, Tanker 71 (N5233V) shows off its natural metal silhouette. This B-17G ultimately joined the Air Force Armament Museum at Eglin Air Force Base, Florida.

Tanker 71 (N5233V) served as a dark blue U.S. Navy PB-1W version of the B-17. By 1963, it was on the books as an air tanker in Medford, Oregon. (Donald Keller Collection)

With all four Wright Cyclone engines safely turning, a fire guard sprints off the ramp with an extinguisher in hand as Tanker 71 prepares to roll. (Donald Keller Collection)

BEST OF THE BEST, 1967

A 1967 air tanker survey published by the Forest Service's Pacific Southwest Forest and Range Experiment Station at Berkeley, California, asked fire air-attack specialists to rate 12 models of fixed-wing and helicopter air tankers for effectiveness in delivering retardant. The survey's authors lamented the lack of realistic and quantifiable testing of air tankers at the time: "In the absence of quantitative data on air tanker effectiveness and because of the small likelihood of obtaining any in the near future, we decided to use an opinion survey technique to rate air tanker effectiveness."[66]

Environmental factors influencing tanker performance included terrain, slope, fire fuel type, wind direction relative to aircraft path, and wind speed. The 11 interviewees ranged from a forest fire control officer to a number of government and civilian pilots. "Fixed-wing tankers as a group were considered more effective than helicopters in operating in various wind speeds and wind directions. They were also rated higher in effectiveness than helitankers in flying over targets on low slopes and easier terrain. However, helitankers excelled on the steeper slopes and rougher terrain. They are, in fact, more effective in canyons, rough terrain, and on steep slopes," the report said. But wind was believed to have a greater detrimental effect on helicopter tankers than on fixed-wing assets.[67]

Fixed-wing tanker types rated in the survey were the Boeing B-17, Consolidated PBY, Grumman AF Guardian, Grumman F7F, Grumman TBM, North American AJ-2, and Northrop F-15 (similar to P-61). The extreme case for the fixed-wing tankers was a fire near the bottom of a narrow canyon in a timbered area with a downwind approach in a wind of 26 to 35 mph. The Grumman AF Guardian, a large single-engine tanker, was rated best in that scenario, with the PBY called least effective. The easiest scenario for the fixed-wing tankers was a grass fire on flat terrain with a weak upwind of no more than 10 mph. "In this situation, the [North American] AJ-2 model tanker was rated highest; the F7F, lowest."[68] Interestingly, the two highest-rated tankers in this survey, the Grumman AF and North American AJ-2, never became mainstays in the air tanker fleet, and the small numbers of these types of air tankers ever campaigned were withdrawn or crashed fairly early in the evolution of air tankers.

Note folding wingtip floats have been removed, rendering this Cat as a landlocked fire bomber. (Kenn Knackstedt)

Grumman F7F. (Author Photo)

Boeing B-17. (Author Photo)

Grumman AF Guardian. (Donald Keller)

Grumman TBM. (USFS)

North American AJ-2. (Avco Lycoming)

The USFS testers summarized their conclusions:

"1. Drops from air tankers of any capacity are hazardous. The tankers used for test all showed that under some conditions a drop on a human can be deadly.

"2. Size of load does not appear to be a major factor in initial impact forces on a human, although accuracy of a drop of large size does not need to be as exact to obtain a hit.

"3. When standing, drops are dangerous at heights under 50 feet. In the prone position the hazard is reduced and is probably safe above 20 feet. The greatest injuries appeared to be caused by the final impact with the ground after being carried or thrown with the liquid.

"4. The dangerous target area was small, particularly lateral to the drop direction. On some drops a hazard existed from loose debris being picked up and thrown with the liquid.

"5. Fatal injuries can be minimized by following these rules:
   a. Lie down facing aircraft with hardhat in place.
   b. Place hand tools off to side.
   c. Grasp something firm to prevent being carried with liquid.
   d. Do not run unless obvious escape is assured.
   e. If in timber stay clear of dead snags, tops, and limbs in drop area. Material such as rocks, live and dead growth, and rolling material on slopes are particularly hazardous."[58]

Corrosion in Air Tankers

Filling aluminum airplanes with any liquid, let alone liquid impregnated with chemicals, is inviting corrosion. Careful painting, cleaning, and draining can keep aircraft relatively corrosion-free, but the air tanker environment warranted a study of corrosion. A 1965 report for the California Air Attack Coordinating Committee summarized post-1964 season inspections of 26 air tankers conducted by fire officials and chemical company specialists. Aircraft are havens for corrosion in places where dissimilar metals touch, where liquids can be trapped, such as skin seams, and where metals like magnesium and fittings like brass control-cable turnbuckles are used. About 25 percent of the tankers inspected were virtually corrosion-free, something attributed to good maintenance procedures. Where corrosion was found, the hardest-hit area was in the retardant tanks of some air tankers. While this might spare the airframe the ravages of corrosion from fire retardant fluids, it nonetheless was a cause for concern. Some pitting on the underside of the wing root of two B-17 air tankers turned out to be similar to pitting observed on a control B-17 inspected that was not an air tanker—corrosion is a universal issue for aircraft operators, not just for those who fly air tankers.[59]

The 26 tankers under inspection most commonly tanked bentonite, Phos-Chek (viscous diammonium phosphate), Fire-Trol (thickened ammonium sulfate), and algin gel. Algin gel appeared to be the most corrosive substance carried by the tankers. The inspectors said the worst corrosion they found was on air tanker E-68, which carried algin gel almost exclusively in 1964. A few other cases of serious tank and gate corrosion were noted in tankers that had carried several loads of algin gel that year. This was in line with USFS corrosion tests from the previous year. One TBM that carried algin gel 95 percent of the time in 1964 developed exfoliation corrosion and pits as deep as 70 mils in its tank and gate. These telltale measurements came as firefighting agencies in California were phasing out algin gel. In addition to recommending scientific choices for materials used in tank construction, the evaluators said: "No corrosion damage was found on exterior surfaces, which were well-painted and maintained in that condition." Epoxy or vinyl resins could inhibit corrosion inside tanks and on drop gates.[60]

Flow patterns visibly left colored retardant stains on the bellies of air tankers and in some protected areas where eddies of slow-moving air could permit thick retardant to build up. The inspectors advised: "Wash the fire retardant residue from all exterior portions of the air tanker as soon as possible upon returning to base after a fire control action. Accumulated retardant will retain moisture or will attract it from the air. This moisture combined with retardant salts causes corrosion." At bases where tankers on standby could be pumped full of retardant quickly, the team recommended parking them empty and filling them only when called.

The 26 tankers inspected during the 1964 California corrosion evaluation represented a wide range of the fleet available at the time:

Tanker Call Sign	*Aircraft Type*	*Tanker Call Sign*	*Aircraft Type*
E-5	N3N-3	E-71	TBM-3
E-17	B-17G	E-72	TBM-3
E-18	B-17G	E-73	TBM-3
E-29	TBM-3	E-75	TBM-3
E-30	AF-2S	E-77	AJ-1
E-34	B-17F	E-78	B-17G
E-35	F-15A	E-88	AJ-1
E-41	F7F-3	E-92	TBM-3
E-42	F7F-3	E-93	PBY-6A
E-46	TBM-3	E-95	TBM-3
E-57	TBM-3	E-96	TBM-3
E-61	B-17G	E-97	TBM-3
E-68	TBM-3	E-98	TBM-3[61]

Qualified Pilots

By 1959, the Forest Service had expectations for the qualifications of pilots who flew air attack missions. A study of southern California air tanker operations that year noted: "In general, pilots who have the background and experience to qualify as air tanker pilots are adaptable to the job of aerial fire fighting. Just meeting the requirements, however, does not mean that a pilot will turn into a top-notch air tanker pilot. The ability of tanker pilots to recognize the target and hit it accurately is as varied as the number of pilots themselves."[62]

More than four decades later, discussions intended to curb air tanker losses touched on a perceived get-by budget attitude held by some in government as well as by some air tanker operators. Some believed there was never enough money to properly operate an air tanker business if increased safety was to be a priority. With this in mind, it is interesting to note comments in a 1959 USFS study of fire operations in Southern California: "Air tanker pilot skill is the result of experience mostly gained on fires and hence at government expense." The report suggested an experienced air tanker pilot should have an airdrop accuracy of about 80 percent. "If pilots, after a reasonable period of time and experience, are accurate approximately 40 to 50 percent of the time, they should not be used until they demonstrate adequacy at their own expense."[63] The report did not quantify what this might cost air tanker pilots.

Chapter Two

ORDER OF BATTLE

Fantastic array of vintage air tankers rests at the end of the day at Wenatchee, Washington, in July 1970 during especially troublesome wildfires in north central Washington State. F7F Tigercat Tanker 41 nearest to camera survived its fire career to earn a home at the Marine Corps museum. Catalina Tanker 49 (N2886D) was a typical U.S. PBY-6A conversion; many Canadian Catalina air tankers were shorter-tail PBY-5A equivalents. PBY Tanker 49 was destroyed in a water crash in 1985 while fighting fires. Ill-fated turboprop B-17F Tanker 34, parked next to the Catalina, did not survive the fire season of 1970. Privateer **Moby Dick**, ***Tanker 42, was consumed by fire on the Wenatchee runway two years after this photo was taken. Barely visible behind the Privateer is B-17 Tanker 10, the*** **Batmobile**, ***which crashed in 1979. A-26 Tanker 74 behind the*** **Batmobile** ***is a survivor of its firefighting career. Tanker number of PV-2 Harpoon at the far end of the line-up is not visible.*** (Author Photo)

California was prescient to the growing dangers of people moving into wildlands. If the concept of an urban-wildland interface was not universal in the 1950s, California was a combustible mix of population and dry terrain that brought the point home, and gave it urgency in the West. During this period, Firestop, a one-year operational study of mass fire behavior and an evaluation of new aids in fire prevention and control, included a June 1955 report, "Aerial Firefighting—Bulk Water Drop by Fixed Wing Aircraft." The Firestop report acknowledged: "Conclusions drawn from five tests indicated that there is little hope of completely extinguishing fires from the air, but it is practical to reduce spread and cool hot spots." The study also summarily dismissed the idea of dropping containers of fluids on fires: "Bombing fires with droppable containers—in fact, any projectile—cannot be tolerated in populated areas or as close support to ground fire fighting forces."[69]

In an era of nuclear attack concerns, the Firestop report also addressed potential civil defense aspects of liquid drops from aircraft: "It may be possible to wash limited targets in fall-out areas by free-fall liquids from fixed-wing aircraft and provide reasonably safe entrance to or egress from places where the radiation level is too high for safe human occupation."[70]

Forest Service officials worked with Willows Flying Service at Willows, California, in 1955 to modify a Stearman PT-17 biplane crop sprayer by placing a drop door at the bottom of its spray hopper. The biplane, uprated with a 450-hp engine for agricultural work, could lift 125 gallons of water or 100 gallons of dense borate mixture. The difference can be explained by the heavier weight of the borate mixture, calculated by Army testers to be 10.5 pounds per gallon as opposed to 8.33 pounds for plain water. Considered the first operational air tanker (if Paul Mantz' TBM of 1954 was still experimental), this PT-17 made its first fire attack on 12 August 1955 over the Mendocino National Forest. Its registration was N75081, Tanker E-1.

Joe Ely of the Forest Service worked with experienced agricultural pilot Floyd Nolta at Willows to develop that pioneer biplane air tanker. Nolta, with three decades of experience at the time, crafted a drop door in the Stearman to release water faster than could be accomplished through typical agricultural spray tubes along the wing. His brother Vance Nolta flew a test drop over the Willows airport on 23 July 1955, extinguishing a set fire in dry grass.

The first record of an organized group of air tankers available to fight fires was in August 1956, when agricultural application aircraft—three N3N and four PT-17 biplanes—were pooled in the vicinity of Willows and Red Bluff, in Northern California. The diminutive biplanes were credited as a factor in gaining control of four-fifths of the 25 fires they attacked that pioneer season. So enthusiastic were foresters about the aerial armada, the biplanes were dispatched hundreds of miles south to battle the Inaja fire in Southern California's Cleveland National Forest in November 1956. Armed with borate slurry, these early air tankers are cited as the

Stearman PT-17
PT-17 (Military)
Length: 24 ft, 9 in
Span: 32 ft, 2 in
Powerplant: 1X 220-hp Continental R-670-5
Empty Weight: 1,931 lbs
Gross Weight: 2,635 lbs
Cruising Speed: 96 mph

Boeing N3N
N3N-3 (Military)
Length: 25 ft, 6 in
Span: 34 ft
Powerplant: 1X 235-hp Wright R-760-2
Empty Weight: 2,090 lbs
Gross Weight: 2,792 lbs
Cruising Speed: 90 mph

Naval Aircraft Factory built the N3N-3 that became an early California air tanker (N45084). Preserved, it was photographed making a demonstration drop in 1981. (Malcolm Gougon Collection)

prototypes that inspired Los Angeles media to coin the term "borate bomber."[71]

Encouraged by the biplanes, the USFS sought larger air tankers and bought eight surplus TBM Avengers from Navy stocks in late 1956. By 1957, some of these were making air drops of 400, and later, 600 gallons. Registration numbers on these USFS Avengers were N-102Z through N-109Z. As California led efforts in the mid-1950s to grow the air tanker fleet, the number of air tankers in that state grew from eight aircraft in 1956 to 26 in 1957. Of these, the State of California began contracting its own air tankers for the first time in 1958, using three N3Ns, four Stearmans, and four TBMs that year. By 1961, an estimated 120 air tankers were in service throughout the United States.[72]

California continues to operate an aggressive air tanker program. Starting in the 1973 fire season, the state operated its own modified Grumman S-2A Tracker air tankers. Statistics as of September 2007 show the California Department of Forestry and Fire Protection's aviation fleet included 23 turbine S-2T Tracker air tankers, 11 UH-1H helicopters, and 14 fast OV-10As used as bird dogs directing air tankers at a fire scene.

Pioneering N3N-3 air tanker as preserved by 1981 conjures early California air tankers. (Malcolm Gougon Collection)

Grumman TBM Avenger

Following Paul Mantz' 1954 demonstrations with a modified TBM-1 Avenger as an air tanker, readily available surplus Avengers were converted to that role. The Avenger could be a mixed blessing. Smaller than multi-engine bombers, it was flown by a solo pilot, meaning fewer souls risked over dangerous fire terrain (and fewer pilots to pay per aircraft), though its typical load of 600 gallons of retardant was eclipsed by larger bombers. Its reliance on a single engine was cause for concern in some firefighting circles, and ultimately this contributed to withdrawal of TBMs from use as air tankers in the United States in the 1970s.

In 1961, the USFS and the U.S. Army's aviation test activity at Edwards Air Force Base, California, evaluated a typical TBM air tanker to better understand its capabilities and limitations. Tanker 08, TBM-3 N108Z, flew the evaluations. Testers determined the optimum retardant drop speed range was between 105 and 120 knots calibrated airspeed. Within that speed envelope, the TBM evaluators noted: "Retardant drops do not induce airframe loads exceeding the structural capabilities of the aircraft." But TBM airframe loads were strongly affected by drop speeds; "structural load limits may be exceeded at drop speeds above 145 knots calibrated airspeed."[73]

The Army and Forest Service evaluators found aspects of the TBM-3 air tanker's handling qualities compromised: "The static longitudinal stability of a modified TBM-3 is weaker than that of a standard model, and to counteract this effect the aft center-of-gravity limit should be 31 percent mean aerodynamic chord instead of the limit of 35 percent mean aerodynamic chord recommended by the Flight Operating Handbook." Removal of military equipment and insertion of air tanker necessities contributed to changed flying qualities, and narrowed the margin for safe aft center-of-gravity limits.

The testers also noted: "The mission assigned to the TBM-3 in its role as an air tanker differs considerably from its design mission. As an air tanker, the aircraft mission is (1) to take off at the maximum gross weight, (2) to fly for a short distance to the drop zone, (3) to perform a drop in close proximity to the ground (i.e., 75 feet off the ground) in mountainous terrain at rather low airspeeds, and (4) to return to the aircraft landing field. The sudden release of 6,000 pounds of chemical solution (or liquid) is considerably different than the dropping of one 1,500-pound torpedo. The airplane is also subject to the severe turbulence associated with the atmosphere of the mountains on hot afternoons and sometimes with the unstable atmosphere generated by a fire. Whenever possible, the bombing runs are planned with a downhill escape route to provide for the malfunction of tank doors and to provide an added margin for error. Occasionally, fighter-type maneuvering is necessary to get out of dangerous situations. Most of the missions are flown in temperatures corresponding to those of an Army Hot Day (a measurement standard) or hotter." In fewer than 200 words the testers summed up the adrenalin-pumping world of early air tanker pilots fraught with danger and unforgiving of errors.[74]

The testers made another observation that cash-strapped air tanker operators no doubt would have found impossible to afford: "The low-altitude mission of the TBM-3 air tanker, coupled with the dependence on a single engine, implies a need for a low-altitude escape system, i.e., a zero-altitude ejection seat."

The stock multi-ribbed TBM front windscreen also came in for its share of criticism from testers who advocated doing what some air tanker operators had already worked out, a one-piece wraparound windscreen. The test TBM had a number of holes and chinks in its structure that allowed carbon monoxide to enter the cockpit. After sealing these gaps, the levels dropped appreciably, although the infusion during high-power settings at takeoff and climb were still objectionable to the testers.[75]

The evaluators described takeoff performance in the TBM air tanker: "Directional control at low speed was adequate, and no difficulty was experienced in holding heading during the entire takeoff run. At a calibrated airspeed of between 72 knots and 77.5 knots, the airplane tended to fly itself off the ground. At lower speeds, the airplane did not become airborne;

TBM-3E (Military)
Length: 40 ft, 11.5 in
Span: 54 ft, 2 in
Powerplant: 1X 1,900-hp Wright R-2600-20
Empty Weight: 10,545 lbs
Gross Weight: 17,895 lbs
Cruising Speed: 147 mph

The rigors of air tanker operations include the possibility of ground looping a heavy tailwheeled aircraft like this TBM-3 Avenger, its right wingtip creased into useless aluminum origami at Columbia, California, summer 1967. Still retaining its famous Grumman-designed wing-folding mechanism, this Avenger takes up less room on the ramp awaiting repairs. (Author Photo)

and holding the aircraft on the ground past 83 knots resulted in a rapid decrease in acceleration and, consequently, a large increase in takeoff distance." The TBM enjoyed an increase in climb performance while turning left, with torque, compared to turning right, against torque. "The best climb speed in a turn was about 10 knots less than the best climb speed in straightforward flight."

The flight mechanics of making a retardant drop with a TBM-3 "consists of a primary pitch-up caused by the drop itself, followed by a mild pitch-down caused by the shift to the forward center of gravity," the test report noted. The recommended airspeed envelope for drops of 105 to 120 knots with 15 degrees of flaps was chosen because slower speeds led to poor visibility due to nose-high attitude and the possibility of a rudder-reversal condition. Speeds higher than 120 knots were disfavored because the airframe sustained ever-higher load factors during release. "The absolute maximum limit for drops, therefore, should be set at a calibrated airspeed of 145 knots. Above this calibrated airspeed during the drops, the limit-design structural load factor is approached, and the abruptness of the maneuver could cause the pilot to lose control of the aircraft." The testers learned from TBM operators: "The technique of leading with nose-down stick just as the release button is pushed used by most of the experienced air tanker pilots reduces the pitch-angle change and improves the pattern the borate makes on the ground."[76]

The Army testers speculated on reasons for the pitch-up phenomenon that are unique to aircraft releasing a large mass of fluid: "The most plausible theory attributes the pitch-up to the blockage or flap effect caused by the wall of liquid the airstream sees at the instant of drop. The blockage causes a low-pressure area aft of this wall (i.e., aft of the center of gravity) and consequently a nose-up pitching moment... This moment would be superimposed on the nose-down moment caused by the forward center-of-gravity shift. In any case, future designers of air tankers should keep this pitching moment in mind because it is the limiting factor in establishing a top speed for the drop envelope."[77]

The testers acknowledged not all TBM air tankers had identical modifications for their role as firefighting aircraft, and weight-and-balance data cited in the report should be re-verified on individual aircraft rather than assuming uniformity throughout the Avenger firefighting fleet.

The testers said many TBM air tankers operated at a forward center of gravity well ahead of the aircraft's intended limit. Lateral control-stick forces of the test tanker were 36 pounds, markedly higher than the 25 pounds desired for horizontal bombers in military specifications. A cut-and-try approach to installing vortex generators on the vertical stabilizer of the test TBM tanker showed some improvements in lateral stability, but the testers acknowledged this was done with "a trial-and-error method with little theoretical background." Hence, not enough conclusive evidence was obtained to warrant fitting all TBM-3 air tankers with vortex generators. "However, there are indications that a properly designed vortex generator system would improve the lateral-directional flying qualities to an acceptable value."[78]

Stalls in the TBM tanker were accompanied by a

Concurrent fueling and tanking are captured in this vintage photograph of a TBM readying for a sortie, pilot at the ready. (USFS)

sharp torque-induced roll to the left. The large rudder could check this roll but only after considerable rolling had taken place. Little warning preceded stalls. To quantify performance of the TBM air tanker, takeoff tests with a full borate load were made at a pressure altitude of 2,800 feet from the Grey Butte, California, airport and at a pressure altitude of 7,000 feet with a load of water hefted from the Long Valley, California, airport.[79]

The test report of the TBM-3 Avenger air tanker characterized the aircraft in the firefighting configuration in a range of observations:

"The TBM-3 aircraft, modified for air tanker use, is capable of accomplishing its intended mission with reasonable safety provided the pilot is familiar with the handling qualities and restrictions of the aircraft.

"The aircraft response during a retardant drop is a controllable maneuver in which the load factor does not

In the blue Montana sky, clouds waft above a worn but colorful TBM Avenger parked in front of one of Johnson Flying Service's remarkable Ford Trimotors. Studies of air tanker types like the TBM reveal variations in tank design. (USFS)

A view few people ever see is an actual fire attack from behind, as a TBM unloads with a desirable downhill escape path ahead of it. (USFS)

Green-labeled bags of powdered retardant slurry mix wait on a pallet as pink-stained TBM Avengers are refueled. (USFS)

Deep-bellied TBM Tanker E26 powers up in summer 1970. Though early single-engine tankers like the TBM fell out of favor, newer turbine-powered agricultural aircraft later re-established the role of single-engine air tankers. (Carl M. Johnsen)

exceed the published limit provided the drop airspeed is within the range recommended in this report.

"Takeoff and climb performance of the TBM-3 are excellent. The optimum climb speed is higher than the Flight Handbook recommends due to the modification of the aircraft."[80]

The testers did not like the directional stability of the TBM-3 air tanker. In uncoordinated rolling maneuvers, the aircraft might pitch up into a stall, they noted. In an unintended compliment to the skilled TBM air tanker pilots who overcame any shortcomings, the testers described what they considered to be less than optimal qualities of their test tanker:

"The TBM-3 exhibits a maneuvering instability at load factors considerably below the maximum allowable maneuvering instability. A forward center-of-gravity location aggravates this characteristic.

"The control forces about all axes of the aircraft are considerably higher than those that are recommended. This deficiency can result in early pilot fatigue.

"No warning is given by this aircraft before an unaccelerated stall. This characteristic creates a potential hazard during low-speed flight at normal drop altitudes and airspeeds."[81]

If the test report gave a mixed review of the TBM-3 as an air tanker, it nonetheless quantified what air tanker operators lived with on a daily basis with their sturdy surplus Avengers. Two decades of TBM air tanker service in the United States and Canada validated the usefulness of the modified Avenger as well as the expertise of its many pilots.

The pioneering use of TBMs in the western United States did not go unnoticed in the South, where fast-growing pine forests are susceptible to persistent wildfires. In January 1958, one of the USFS 440-gallon TBMs was loaned to the Forest Service's Southeastern Forest Experiment Station for testing and training involving North Carolina and Georgia state forestry organizations. For calibration drops on a measuring grid, the TBM carried a slurry mixed with 1,250 pounds of kaolin clay and enough water to produce a 440-gallon load. Kaolin, local to Georgia and other parts of the South, is a fine clay substance used for coating slick publishing papers. For the 1958 tests, its local availability no doubt contributed to its favorable price over that of borate, mined in the southwestern United States. Kaolin slurry has a consistency similar to that of borate, and it clings well to vegetation. Even brighter white than borate, kaolin is easier to see following an airdrop.[82]

Following calibration with inexpensive kaolin, drops on test fires were made with borate solution and wet water (water treated with a wetting agent). Both were effective, although foresters noted the variety of undergrowth and overgrowth in the South caused variations in coverage reaching the ground.[83]

Canopy open, the pilot of TBM Tanker A12 brings his Avenger in for a throttled-back landing at Coeur d'Alene, Idaho, during the Clark Fork fire in the northern part of the state in July 1973. (Author Photo)

North American B-25 Mitchell

In the late 1950s as air tanker operations gained momentum, a ready supply of recently retired Air Force B-25 Mitchell medium bombers seemed natural for the job. Already stripped of armament, these post-war B-25s finished their military careers in several utility and training roles.

Two 1960 B-25 air tanker crashes (one in Oregon and one in Washington) put an early end to the use of Mitchells in California firefighting operations and led to Forest Service and Army tests of a TB-25N tanker the following year.[84] The USFS report on the tested air tanker B-25 was to the point: "The tested TB-25N airplane modified for use as an airtanker is not capable of accomplishing the airtanker mission with reasonable efficiency and safety. The airplane is not stressed to withstand the loads imposed on it by dropping liquid chemicals from the bomb bay unless the load dropped is in increments of 3,000 pounds or less. Also, airspeed must be kept within a narrow recommended range and only moderate simultaneous turbulence or maneuvering loads can be tolerated."[85] Yet the wartime B-25 Mitchell bomber was a combat hero, and years after these tests, vocal advocates remain devoted to the B-25 in its many iterations.

For the tests, the Forest Service bought TB-25N air tanker N10564. It operated with a 1,240-gallon tank in the bomb bay that could be emptied in one operation, or toggled between front and back compartments with a delay in release of the second compartment. The testers immediately noted the small size of the crew hatch in the belly of the bomber, just ahead of the bomb bay. The hatch opening and other constraints would make bailout difficult, especially if the aircraft were in uncontrollable flight, the report observed. "The pilot and copilot positions should be supplied with ejection seats so that escape would be possible at low altitude," the report optimistically said, seemingly indifferent to the fiscal realities of air tanker operations of the day.[86]

On the test TB-25N, the cowl flap levers on the corner of the control pedestal were modified to become door release handles for the retardant tank. The report explained: "Levers to control the cowl flaps have been installed behind and above the copilot's head. This is an undesirable location for these controls." The testers explained a factor they believed relevant to aircraft, including air tankers, subjected to frequent varying load factors: "The design limit load factor of an airplane is that load which the aircraft structure should be capable of withstanding without yielding, i.e., receiving a permanent deformation. The ultimate load factor is the maximum load an aircraft structure can sustain without structural failure."[87]

The ability to quantify a particular airframe's load factors was questioned by the testers: "None of the preceding factors can be guaranteed in all aircraft structures. Throughout the years an aircraft remains in service, it is subjected to conditions, which weaken the structure: corrosion, structural modifications, turbulence, hard landings, vibrations, and general hard usage. These are variables that cannot be accounted for in the original determination of the structural strength. No guarantee exists that the design limit load factor, margin of safety, or ultimate load factor have not been compromised or changed." Fuselage loading—in this case, the bomb bay tanks—contributed to wing bending moments, the study noted.[88]

For airdrop tests, the Forest Service's TB-25N performed most of its releases at an altitude of 7,500 feet in smooth air. Four drops were made at 75 feet above ground level (AGL) at an altitude of about 2,300 feet at the Grey Butte, California, airfield southeast of Edwards Air Force Base. Drops included borate solutions and one less-viscous 600-gallon water release. In anticipation of altitude gain immediately following the drop, "the pilot anticipated the drop with nose-down control," the report said. The copilot activated the drop controls. Use of 15 degrees of wing flaps moderated accelerations on the airframe at the time of release. "The use of more than 15 degrees of flap is not recommended, however, since unknown torsional stresses are placed on the wing structure when the aircraft is rotated around its lateral axis (running through the center of gravity from wingtip to wingtip) with the flaps down."[89]

At lighter loads, the use of thicker borate solution decreased load factors from those experienced with dropping water. When 1,000 gallons of borate was dropped, testers determined it took longer to evacuate

The deep belly tank of B-25J Tanker C39 was larger than tanks used on other B-25s. This B-25 retains its individual exhaust stacks manifested as bumps on the engine cowling. Many B-25s were retrofitted with semi-collector rings eliminating some of the individual stacks. (USFS)

the bomb bay tanks than water would: "Since borate is more viscous than water, more time is required for it to drop out of the tanks into the airstream. As a result, approximately one-half second more time is required for the aircraft to reach its condition of maximum cg normal acceleration."[90]

Motion picture footage of air tanker drops sometimes reveals pitching corrections made by the pilot. For the test TB-25N, the Forest Service and Army evaluators noted a moderation of this happened when two 300-gallon releases were made with half-second intervals, versus one release of 600 gallons, especially with water. Less moderation was noted when thicker borate was released. Test equipment at Grey Butte showed the TB-25N gained 50 feet in altitude with the release of 300 gallons of borate; similar releases of water caused the bomber to gain 60 feet in altitude. When 500 gallons of borate were released at once, the TB-25N rose 90 feet.[91]

Rolling pullout maneuvering during retardant release, sometimes employed if terrain dictated, added stress to the B-25. "A rolling pullout maneuver imposes the most extreme loads on an aircraft structure. It is highly probable that under conditions of heavy turbulence, when dropping over 3,000 pounds...and in a rolling pullout, a TB-25N could easily exceed the design limit load factor and incur structural damage as a result." The test report said drop speeds should be confined to an envelope of 105 to 145 knots calibrated airspeed (CAS). This put the B-25 just above its single-engine minimum control speed, and kept it below speeds that aggravated airframe loads. Yet this recipe confronted potential problems from turbulence: "...the recommended drop regime (105 to 145 knots CAS) determined from the tests falls in the caution area for penetrating average turbulence. Operating in areas of high gusts at airspeeds within the chemical drop range would subject the aircraft to dangerous loads."[92]

The evaluation of mechanical forces affecting B-25 air tankers sheds interesting light on the world in which first-generation air tanker operators flew. Even after B-25s were withdrawn from use as air tankers in the lower 48 states, Alaskan operators used B-25s as late as 1973; evidently not everyone agreed they were unsuited to air tanker operations.[93] Some B-25 air tankers languished in the 1960s following their withdrawal from service, only to rise again as part of the remarkable Air Force movie resurrected by Frank Tallman for use in the motion picture *Catch-22.*

B-25J (Military)
Length: 52 ft, 11 in
Span: 67 ft, 7 in
Powerplant: 2X 1,700-hp Wright R-2600-92
Empty Weight: 19,480 lbs
Gross Weight: 35,000 lbs
Cruising Speed: 230 mph

The U.S. Forest Service worked with U.S. Army flight testers in California to characterize the traits of B-25J N10564 as an air tanker. Though serving for many years in Alaska and Canada, B-25s were not long on the air tanker roster in the lower 48 states. Vent is visible atop fuselage over bomb bay to promote quick, easy departure of retardant without "Coke bottling," or splashing unevenly as fluid does from an unvented container. (AFFTC History Office)

With unused gun windows painted over, B-25J, N8196H, earned its keep as an Alaska air tanker. Its retardant load is carried in a flush-mounted bomb bay tank. (USFS)

Boeing B-17 Flying Fortress

If the current population of surviving B-17 Flying Fortresses owes much to the air tanker industry that plucked numerous Fortresses from the scrapper, the rigors of aerial firefighting also claimed a number of Fortresses and their crews. The migration toward larger multi-engine air tankers brought B-17s into this field beginning in 1960. That year, Abe Sellards converted B-17F N17W to air tanker mode, and operated it on a fire in August. Simultaneously, Fast Way Air Service put B-17G, N3702G, into air tanker service out of Chino, California, the same month.[94]

Retardant tank capacity in converted B-17s was typically between 2,000 and 2,400 gallons, with rated capacity listed as 2,000 gallons. Following discovery of a structural crack in a B-17 air tanker in 1973, the Federal Aviation Administration (FAA) limited the retardant load that could be carried in B-17s to 1,800 gallons. By 1980, the USFS was distancing itself from the Fortress, and surviving air tankers segued into museums and flying warbird collections.

The first downed Flying Fortress air tanker was B-17G N3678G, Tanker 15, reported crashing into a mountain near Fallon, Nevada, on 16 October 1962. It would not be the last. Beloved as the B-17 is in the pantheon of great aircraft, Fortresses nonetheless suffered significant attrition as firefighters. The roster of B-17 air tankers lost includes nine bombers; more than one-third of the 23 Fortresses converted as firefighters:[95]

- B-17G N3678G (Tanker 15) crashed in Nevada 1962.
- B-17G N7228C (Tanker B31) destroyed in 1967.
- B-17F N1340N (Tanker A34, E34, 35) crashed in Wyoming 1970.
- B-17G N9324Z (Tanker E18, 18) crashed in 1971; major portions may be restored.
- B-17G N73648 (Tanker B11, E56) crashed in New Mexico 1972.
- B-17G N620L (Tanker 54, C54) crashed in Nevada 1973.
- B-17G N621L (Tanker 64, C64) destroyed in 1975.
- B-17G N66573 (Tanker 10, A10, B10, 33, E85) crashed in Montana 1979.
- B-17G N6694C (Tanker C12) crashed in 1980; some parts saved for restoration.

These losses are not an indictment of B-17s or their crews, but rather a testimonial to the rigors of air tanker operations in the Fortress era.

One crash in particular captured the interest of fliers and enthusiasts alike: B-17F N1340N (Tanker 34) was like no other Flying Fortress. Its four R-1820 radial engines and Hamilton-Standard three-blade propellers were switched for pencil-thin Rolls-Royce Dart turboprops from a Viscount airliner in 1970. Swinging square-tipped four-blade props, the long and light Dart engines were mounted far forward of the original propeller line of the B-17, with sheetmetal fairings accommodating the vast difference in diameter of the Darts from the original Wright engines. The

B-17G (Military)
Length: 74 ft, 4 in
Span: 103 ft, 9 in
Powerplant: 4X 1,200-hp Wright R-1820-97
Empty Weight: 36,135 lbs
Gross Weight: 65,500 lbs
Cruising Speed: 182 mph

Flying with Aero Enterprises out of Fresno, California, when photographed on 9 April 1966, Tanker 34 (N1340N) was a conventionally engined B-17F. Four years later, it would be the only turbine-powered Fortress air tanker. (Kenneth G. Johnsen)

available turboprop power was so great it enabled the crew to return from a fire drop, with the bomber empty, using only the two inboard engines. Aero-Flite of Cody, Wyoming, operated N1340N during the summer of 1970, and the future looked good for this remarkable upgraded air tanker. But it all ended abruptly on a fire site near Dubois, Wyoming, on 18 August 1970 with the fatal crash of N1340N.

The National Transportation Safety Board (NTSB) called the accident a stall (mush) mishap in which the bomber crashed into trees during the pull-up after making a drop run. A high-density altitude may have contributed to the crash as the bomber made a downslope, downwind fire run and pull-up, according to the NTSB report.[96] Stories have circulated that superheated air over the fire robbed the propjets of their power at a critical moment, but this is disputed. Both pilot and copilot perished in the wreck, giving their lives in the ongoing battle against wildfires.

Rugged ribbing of B-17 tank drop doors shows strength built into gates that had to hold the weight of thousands of pounds of liquid retardant until released over a fire. (USFS)

B-17 Tanker 10, dubbed the **Batmobile** ***for its scalloped paint trim, receives servicing under heavy skies at Wenatchee, Washington, in July 1970. Conflagrations in north central Washington caused a hastily convened congress of air tankers to descend on Wenatchee that summer. Tanker 10 was written off in a crash in 1979.*** (Author Photo)

N66573, B-17 Tanker C10, flew with a pinched-down tail fairing at the aft end of the fuselage in place of the original gun emplacement. This venerable Fortress crashed in 1979. (USFS)

B-17 Tanker 78 rested with its four drop gates hanging open. One of the first two B-17s converted to air tanker configuration; this aircraft (N3703G) saw service in 1960. It subsequently joined David Tallichet's massive warbird collection, and flew as the **Memphis Belle** ***in the 1989 movie of the same name.*** (Peter M. Bowers Collection)

The high-pitched whirring of turbine engines accompanies the loaded takeoff of Tanker 34, a one-off modified B-17F, at Pangborn Field in Wenatchee, Washington, in July 1970. Retardant tank protrudes below fuselage contours without streamlined fairing. (Author Photo)

Paused at Wenatchee's Pangborn Field during a lull in July 1970 fire activity, B-17F Tanker 34 shows its unorthodox turbine engine installations. Original nacelles are necked down to match the slim diameter of Dart turboprop engines. Outboard turbine exhaust deflects down and to the rear; inboard engine exhaust angles down, back, and outboard to avoid the main tire, which rests exposed in the nacelle. Long Dart engines put propeller line many feet ahead of that on a standard Fortress.

Turboprop B-17 air tanker N1340N leaps into the air loaded for fire at Wenatchee, Washington, in July 1970. Bombardier's nose section and panels on wings, tail, and aft fuselage are bright orange against natural metal finish. (Kenneth G. Johnsen)

Form follows function in one of the few aesthetic angles on the sole re-engined turboprop B-17F air tanker, N1340N. Rolls-Royce Dart turboprop engines provided more than enough power for the air tanker. (Author Photo)

Consolidated PB4Y-2 Privateer

The large PB4Y-2 Privateer, with a wingspan of 110 feet and the capacity to tank 2,000 gallons in its two-part bomb bay, came on line as an air tanker as far back as 1959. The first of the large four-engine surplus bombers to become air tankers, Privateers outlasted nearly all of their World War II contemporaries by the time the last PB4Y-2 Super Privateers were withdrawn from service following a crash in 2002. Initial users of Privateers retained the bombers' standard Pratt & Whitney R-1830-94 normally aspirated (non-turbosupercharged) engines housed in characteristically oval cowlings. Around 1970, many Privateers began undergoing modification to use four 1,700-hp Wright R-2600-92 engines and cowlings from B-25 Mitchell bombers. An abundance of R-2600s made the swap feasible. Although the Privateers were still redlined at their earlier performance parameters, the availability of the extra power was probably a comfort to pilots nonetheless. Resulting aircraft were identified as Super Privateers.

By 1964, one Privateer operator had responded to a common lament of pilots of Privateers and earlier similar B-24s: Numerous crossmembers in the factory-built cockpit canopy and windscreen blocked visibility. This could be particularly dangerous over a smoky fire scene, with air tankers and bird dogs sharing busy airspace. Tanker 85, N3191G, had its multi-paned cockpit canopy replaced with a simple enclosure made up of a few flat panels affording greater visibility. Later, Hawkins and Powers Aviation in Greybull, Wyoming, refitted their fleet of Privateers with curved canopies that offered fewer bars to block visibility than on a standard PB4Y-2. The ability to look forward through the nose of a Privateer from the flight deck was a benefit to maneuvering air tanker pilots. Some air tankers were ex-U.S. Coast Guard variants that featured bulbous Plexiglas nose enclosures that enabled forward visibility. The solution for other Privateer air tankers came in the form of an early B-24D multi-pane Plexiglas nose mated to Privateer Tanker C30, N3739G, and upended B-47 Stratojet canopies faired into the noses of some Hawkins and Powers Super Privateers.

Not all air tanker mishaps result in fatalities. Tanker 85, PB4Y-2 N3191G, came to rest in about 8 feet of water in Oregon's Diamond Lake on 27 July 1970, attributable to fuel exhaustion, according to a National Transportation Safety Board (NTSB) report.[97] Remarkably intact, but with no easy way out of the mountain lake, Tanker 85 succumbed to the scrapper.

Attrition, retirement, and buyouts dwindled the number of Privateer air tanker operators until only the largest, Hawkins and Powers Aviation, remained. Tanker 123, N7620C, crashed 18 July 2002 while fighting a fire in Colorado when its left wing came off the aircraft. In December 2002 the USFS and the Bureau of Land Management (BLM) decreed Privateers would no

PB4Y-2 (Military)
Length: 74 ft, 7 in
Span: 110 ft
Powerplant: 4X 1,350-hp Pratt & Whitney R-1830-94
Empty Weight: 37,485 lbs
Gross Weight: 65,000 lbs
Cruising Speed: 140 mph

PB4Y-2 Privateer 59818 departed the Navy with its unarmed turrets intact. Its conversion to an air tanker was about to begin as this photo was taken in Arizona in the early 1960s. This bomber became Tanker 30, registered N3739G, and outlived its fire career as a museum piece. (Ken Shake)

longer be used as federal air tankers.[98] This ended an impressive run of Privateer operations spanning 43 years. For much of that time, Hawkins and Powers Aviation was synonymous with Privateer air tankers, nurturing the big World War II PB4Y-2s long after the retirement of B-17 air tankers, adding evidence that Privateers may have been the better bomber when it came to firefighting.

Sold at auction, the remainder of the Hawkins and Powers Privateers are destined to become restored warbirds or museum pieces.

Privateer air tanker losses include:

- PB4Y-2 N6884C (first Privateer with this N-number), demolished in off-airport landing near Prescott, Arizona, June 1959.
- PB4Y-2 N3191G (Tanker 85), ditched in Diamond Lake, Oregon, 27 July 1970.
- PB4Y-2 N6816D (Tanker 42), inflight fire consumed aircraft on landing, Wenatchee, Washington, 27 July 1972.
- PB4Y-2 N7237C (Tanker 50), crashed near Safford, Arizona, 27 June 1974.
- PB4Y-2 N6813D (Tanker 125), ditched off Vancouver Island, B.C., Canada, in 1975.
- PB4Y-2 N2870G (Tanker 122), damaged and written off at Ramona, California, July 1980.
- PB4Y-2 N7620C (Tanker 123), crashed near Estes Park, Colorado, 18 July 2002.

Blunt rear fuselage on Privateer air tanker creates air currents favorable to plastering the aircraft with soupy fire retardant. The same qualities of retardant that cause it to cling tenaciously to vegetation can adhere it to aircraft until hosed off at the end of the day as a corrosion-control measure. (Author Photo)

Three of Hawkins and Powers' fleet of Super Privateers (123, 126, and 121) are deployed to Fox Field in California in this image that also records the presence of a C-54 called in to battle wildfires. The ramp at Fox Field's air tanker base holds large numbers of tankers called in during fire operations; fewer once the danger subsides. (Author Photo)

Tanker 127 lays down the red, white, and blue in a compartmentalized demonstration drop at the National Championship Air Races in Reno, Nevada, in September 1975. More than a stunt, the ability to drop separate tanks in trail allows the Privateer to lay a longer continuous line of retardant on a fire line. (Author Photo)

Hawkins and Powers' PB4Y-2 Tanker 123 climbs away from Fox Field, Lancaster, California, its four 1,700-hp Wright R-2600-92 engines laying smudges from their B-25 exhaust stacks. Retardant from previous sorties stains the belly of the bomber, to be washed off at day's end.

Rosenbalm Aviation's Tanker 85, photographed at Medford, Oregon, on 13 April 1966, showed an early example of enhanced cockpit canopy visibility. Tanker 85 featured an angular specially made canopy with fewer obstructing webs than a standard Privateer/ Liberator canopy. (Author Photo)

The classiest nose job given to a Privateer air tanker is this B-24D greenhouse mounted to the front of Tanker 30 (N3739G) of T&G Aviation photographed at Chandler, Arizona, in January 1980. Privateer tanker operators sought better forward and downward visibility for aircrews. (Author Photo)

Test drop in Arizona shows Privateer pattern with aircraft N3739G before it received its Tanker 30 number, circa 1963. Removal of tracked roll-up bomb bay doors from the Privateer enabled installation of retardant tanks with self-closing gates. (Ken Shake)

PB4Y-2 SUPER PRIVATEER WALKAROUND PHOTO SERIES

Every modeler's dream is to be able to walk around a particular airplane while closely observing markings and structural details to be faithfully replicated in a model. That is what was done with this Consolidated PB4Y-2.

Exterior views of PB4Y-2 Tanker 124 of Hawkins and Powers Aviation, photographed at Boise, Idaho, in 1989, provide details about a working Privateer air tanker of that era. Dark area extending on belly from retardant tanks to aft fuselage helps protect skin from corrosion. Vastly simplified cockpit canopy provides much greater unobstructed field of view for pilots. (Author Photo)

124

N2872G
124

124

N2872G

Engine packages from B-25 Mitchell medium bombers replaced Pratt & Whitney R-1830 engines on most Privateer air tankers. With semi-collector ring exhaust, the multiple exhaust ports of the B-25 smudge the wing and cowl flaps. (Author Photo)

Deep wheel well in right wing of Privateer leaves the tire exposed when retracted. Wing structure can be discerned inside the well. (Author Photo)

Pilot's seat of Privateer Tanker 124 shows complex structure to permit adjustments. Worn sunshade overhead lets pinholes of daylight stream in. Inward-hinging hatch at left of photo is used by crew to enter and exit Privateer via bolted-on exterior ladder. (Author Photo)

Instrument panel of Privateer Tanker 124 shows air tanker layout. (Author Photo)

Periodic wash-downs keep the Privateer clean and help prevent corrosion. (Author Photo)

Copilot's side view shows typical Convair Z-stringer construction in fuselage. (Author Photo)

Consolidated PBY Catalina

Slow and steady, the 1930s-era PBY Catalina, with its semi-circular fuselage cross-section and wide V-bottom boat hull, became a reliable air tanker in the United States and Canada (see Chapter Three). Typically, Catalinas converted to air tankers in the United States were late-model tall-tail PBY-6A amphibians, while those in Canada tended to be PBY-5A or equivalent Canadian-built Canso amphibians. Some U.S. Catalinas were relegated to land operations, and had their characteristic folding wingtip floats removed. Others, including pioneering Cansos in Canada, retained water-operating capabilities and used retractable probes to scoop water from lakes for firefighting. Catalinas could tank as much as 1,500 gallons.

PBY-5A (Military)
Length: 63 ft, 10 in
Span: 104 ft
Powerplant: 2X 1,200-hp Pratt & Whitney R-1830-92
Empty Weight: 20,910 lbs
Gross Weight: 35,420 lbs
Cruising Speed: 113 mph

Re-engined Super Cat PBY-6A Tanker 85 (N6453C) retained its original search side blisters, which could be rolled up and opened. (Malcolm Gougon Collection)

Tanker E49, PBY-6A Catalina N2886D, taxied during fire operations at Wenatchee, Washington, in July 1970. Catalina air tankers often operated as landplanes. (Author Photo)

Super Catalina Tanker 49 used an add-on external boarding ladder. (Donald Keller)

Side view of PBY-6A Tanker 85 (N6453C) shows altered contours of upper nose. This so-called clipper bow faired over the Catalina's traditional nose gun emplacement. Open drop door can be seen inside and ahead of mainwheel. (Author Photo)

Like stout manhole covers, retardant-smeared drop doors on a PBY Catalina amphibian were designed to withstand the rigors of water landings. (USFS)

PBY-6A Tanker E54, powered by stock R-1830 engines, flew for Cal-Nat Airways when photographed at Columbia, California, in August 1967. (Author Photo)

de Havilland DHC-2 Beaver

The USFS used air tanker modifications of Canadian de Havilland DHC-2 Beaver high-wing, single-engine bush planes over forests of the northern Midwest (see Chapter Three). More than 970 DHC-2 Beavers were delivered to the United States Army and Air Force between 1950 and 1960. Many others served civil operators in the bush environment.

Historically, in 1957 a 125-gallon tank was fitted inside the cabin of a Noorduyn Norseman single-engine bush plane. Water was dumped through a 23-inch hatch in the aft cabin floor. The following year, scooping equipment was fitted to a Beaver, which released water from a 17-inch-diameter floor hatch. While showing the way, these modifications were less than ideal as the tank obstructed the cabin and water evacuation speed was slower than desired. In that era Canadian engineers developed a pair of cylindrical tanks attached to the tops of the twin floats and fed by scoop tubes. These tanks were emptied by rolling them over, with a wind vane turning the tanks back to the upright position for the next loading. A USFS report cited some items on this system they wanted to improve: "Although this design has the advantage of rapid water release, a divided splash pattern was made which was less effective than the single more highly concentrated pattern from a single tank. The presence of tanks prevent(s) the use of the pontoon racks for cargo dropping, and the trunnions prevent adequate streamlining of the tanks. Thus, seaplane characteristics were affected in that there is a 12 mph loss of cruising speed."[99]

By early 1962, the Forest Service's Ely, Minnesota, Service Center had perfected an external water tank for use on floatplane variants of the Beaver. Using a surplus Fletcher 230-gallon aircraft drop tank cut down to 125 gallons to meet Beaver weight limits, the design freed up space in the Beaver's cabin and floats that had been consumed with previous tank configurations. It also yielded a better drop pattern than some previous twin-tank iterations. The streamlined tank fit under the fuselage and between the floats, in space previously unused. With

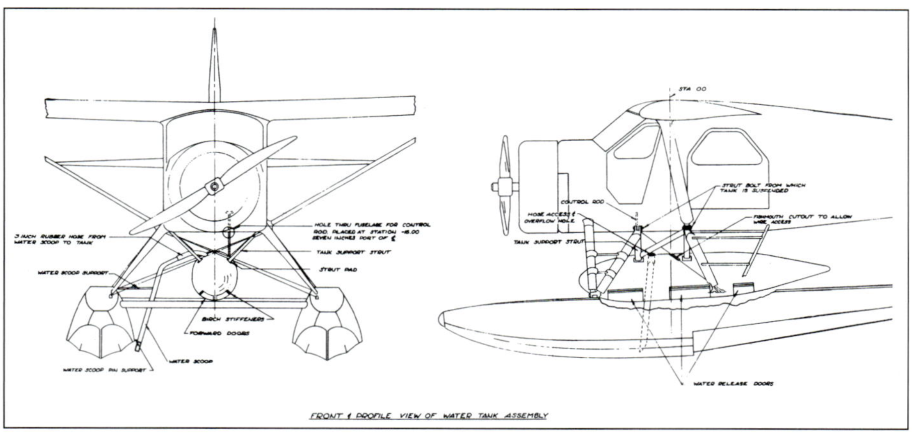

U.S. Forest Service drawings depict a modified drop tank installation on a float-equipped de Havilland Beaver, making the floatplane a small-scale scooper.

some modification, the installation could also be made on the similar de Havilland Otter floatplane. "There is a notable decrease in air drag and cabin space and pontoons are available for other uses," the Forest Service report documented. "Drop patterns are greatly improved, and the tank is easily mounted for almost instant use; it takes only a few minutes to install the pickup tube."[100]

The tank was filled in about 10 seconds from a scoop tube as the Beaver taxied at about 40 mph on the water's surface. Once filled, excess water escaped through an overflow vent on the top of the tank. The scoop tube was mounted to the right float at a height that kept the tube above water level at takeoff speed. The tank was fitted with three sets of double drop doors giving a total opening of more than 700 square inches. A mechanical control rod in the cockpit, pushed down by the pilot, opened the doors. The doors could be reset in the closed position by pulling up on the rod.[101]

de Havilland DHC-2 Beaver (Military)
Length: 30 ft, 3 in
Span: 48 ft
Powerplant: 1X Pratt & Whitney R-985
Empty Weight: 2,850 lbs
Gross Weight: 5,100 lbs
Cruising Speed: 143 mph

Douglas A-26 Invader

The Douglas A-26 Invader, a lean, forward-thinking product of the design team headed by Ed Heinemann, participated in World War II, Korea, Vietnam, and other cold war flare-ups. It served the United States and numerous foreign countries as a fast attack bomber. The availability of surplus A-26 Invaders made them prime for use as early-iteration air tankers. Tanking 1,200 gallons, Invaders were a common element in the air tanker fabric of the United States and Canada from the 1960s. References are made to the first use of an A-26 air tanker in 1960. Into the 1980s, their presence in American fire operations dwindled.

For some, the A-26 will be inextricably linked to the romance of aerial firefighters as portrayed in the Steven Spielberg motion picture *Always,* where pilots including hotshot Richard Dreyfuss and fledgling flier Holly Hunter took A-26s into the wildfire furnace in dramatic depictions of air tanker operations.

A-26C (Military)
Length: 51 ft, 3 in
Span: 70 ft
Powerplant: 2X R-2800
Empty Weight: 22,850 lbs
Gross Weight: 35,000 lbs
Cruising Speed: 284 mph

"Flaps down and gates open" as an A-26 tanker conversion makes a test drop in Arizona. Tanker testers quickly established optimum drop speeds; often slow enough to warrant using flaps. (Ken Shake)

Solid-nose A-26 Tanker 12 (N5544V) was one of two on duty in Tennessee summer 1965. (Author Photo)

Extended flaps on A-26 Tanker 56 at Troutdale, Oregon, show the airflow-directing segments that tamed the hot bomber for landing. (Author Photo)

World War II Douglas A-26 Invader, later called B-26, served as a twin-engine air tanker for many years. Tanker 55 served with Aero Union when photographed at Salt Lake City in September 1971. (Author Photo)

Chin hatch in nose of A-26 Tanker F29 provides crew access. A-26 tankers flew with a variety of Plexiglas bombardier noses (this one has its upper surfaces painted) or aluminum noses originally intended for strafing with ports for machine guns. (Carl M. Johnsen)

Looking aft at the bomb bay drop gates of a Douglas A-26 air tanker, the bomber's large nosewheel door, with built-in step, can be discerned at the right of the image. (USFS)

Douglas B-18 Bolo

At least two models of the stout twin-engine prewar Douglas B-18 Bolo bomber were tanked for firefighting. Tanker 120, an older blunt-nosed B-18 model, registered N52056, ended its flying days as an air tanker with Hawkins and Powers in Greybull, Wyoming. This bomber later became part of the collection of the Castle Air Museum on the site of the former Castle Air Force Base, Atwater, California. A slightly newer B-18B, modified during World War II from a B-18A and subsequently given civil registration N66267, flew with Roberts Aircraft of Boise, Idaho.

B-18 Specifications
Length: 56 ft, 8 in
Span: 89 ft, 6 in
Powerplant: 2X R-1820-45
Empty Weight: 15,719 lbs
Gross Weight: 21,130 lbs
Cruising Speed: 167 mph

Aiming for oil drums, the pilot of this rare Douglas B-18 tanker demonstrates the bomber's capabilities. (USFS)

Douglas DC-3/C-47

C-47 (Military)
Length: 63 ft, 9 in
Span: 95 ft, 6 in
Powerplant: 2X R-1830-92
Empty Weight: 18,200 lbs
Gross Weight: 26,000 lbs
Cruising Speed: 135 mph

Perennial airliner, transport, movie star, and icon, the Douglas DC-3 of the 1930s saw limited use as an air tanker four decades later. Three DC-3C variants were modified to be able to drop retardant, although only one of these flew on fires. With a capacity of 875 gallons, the twin-engine Douglas DC-3 air tanker used the same tanks it also carried for area spraying, fitted with dump valves to permit the aircraft to release retardant at the pilot's command. More than one operator fitted DC-3s to become air tankers in the 1960s, but the only one used saw service for three seasons starting in 1970.[102]

Tanker 40, a low-slung PV-2 Harpoon dating back to World War II, settled in at Wenatchee, Washington, after a sortie in July 1970. (Author Photo)

Douglas C-54 (DC-4) Skymaster

Douglas Aircraft's archetypal four-engine transport, the C-54, figured prominently in military operations including carrying the Japanese surrender delegation from the small island of Ie Shima to the Philippines in August 1945, as well as carrying huge amounts of lifesaving supplies into Germany during the Berlin Airlift several years later. (At least one Aero Union C-54 air tanker was said to contain remnants of

C-54 (Military)
Length: 93 ft, 10 in
Span: 117 ft, 6 in
Powerplant: 4X R-2000-7
Empty Weight: 37,000 lbs
Gross Weight: 62,000 lbs
Cruising Speed: 178-200 knots

Tanker 19, a C-54, showed its bulging retardant tank during takeoff from Fox Field, California, in May 1994. Unlike aircraft built with internal bomb bays, transports like the C-54 relied on external tanks to make them viable firefighters. This was simpler than redesigning the transport fuselage to accommodate a large reinforced opening for drop gates. (Author Photo)

coal dust attributed to its use as a coal-hauler in the Berlin Airlift.)

The availability of military surplus C-54s and some commercial DC-4 airliner variants yielded air tankers with retardant capacities ranging from 2,000 to 2,200 gallons. So-called "Super DC-4" versions were modified with more powerful engines enabling the higher retardant load and faster speeds.[103]

As C-54 Tanker 19 settles in at Fox Field circa 1996, the history of these Douglas transports rides with it. C-54s became the icons of the Berlin Airlift; a C-54 carried the Japanese surrender delegation to Manila to plan the end of World War II; and many C-54s finished lengthy careers as air tankers. (Author Photo)

Douglas DC-6

Douglas DC-6 airliners and similar C-118 military variants were converted to become air tankers capable of delivering 2,450 gallons of retardant.[104] Two elegant DC-6B piston-engine airliners were instrumented for a NASA evaluation in the early 1970s. The outcome of the NASA project indicated pilots sometimes underestimated the flight loads their maneuvering imposed on the DC-6 air tankers (and by presumption, other air tankers as well). The design maneuver limit load—the maximum an aircraft is expected to encounter in service—was equaled or breached by the instrumented DC-6 air tankers in 10 percent of their airdrop recovery maneuvering. On one DC-6 tanker the ultimate load factor—usually not attained in service—was exceeded when the tanker pilot had to

Air tanker operators and user agencies have studied the way fluid falls from an airplane for decades. This photo depicts a 1977 demonstration drop by a DC-6 in Canada. (Author Photo)

maneuver to avoid a canyon wall obscured by smoke. The aircraft did not exhibit any structural damage attributed to its light weight following drops on the fire. The NASA report observed: "The severity of maneuver load applications, in both magnitude and frequency of occurrence, is such that significant shortening of the structural life of the aircraft should be expected." Pullout load factors on the instrumented DC-6s during retardant drops and recoveries ranged from -0.5 g to +3.9 g.[105]

DC-6
(adapted from C-118A military specifications)
Length: 106 ft, 10 in
Span: 117 ft, 6 in
Powerplant: 4X R-2800-52W
Empty Weight: 55,357 lbs
Gross Weight: 107,000 lbs
Cruising Speed: 307 mph

Douglas DC-7

DC-7 airliners with long ventral tanks attached to the fuselage were certified for carrying 3,000 gallons of fire retardant. A quick recognition feature for the DC-7 to distinguish it from the similar DC-6 is the use of four-blade propellers on the DC-7.[106]

Aero Union Corporation of Chico, California, designed the AU3100-H retardant tank for external mounting to the bellies of DC-6s and DC-7s. Capable of holding 3,000 gallons, this tank has eight compartments and eight individual doors that can be triggered to provide customized drop patterns, either manually or automatically sequenced. A full salvo from this heavyweight tank can cover a swath 80 feet wide and 300 feet long; by activating the eight doors singly in trail, a swath 40 feet wide and 2,400 feet long can be achieved. The tank uses two rows of four doors in tandem.[107]

The left-hand grip of the pilot's control yoke houses a guarded pushbutton that triggers the door operating mechanism as programmed on a drop-selector dial. The specter of an inflight emergency requiring immediate jettisoning of the entire retardant load is addressed by an emergency dump control knob mounted just below the instrument panel glare shield where it can quickly be accessed by either the pilot or copilot. In normal operations, the retardant tank door system relies on the aircraft's hydraulic system. An emergency fluid and air accumulator is available to power the doors via the dump knob. It must be charged and under pressure to function. This is the only way to open the tank doors in an emergency.[108]

The tank bays have one-way check valves that aid in filling the tank. As a result, normal operations call for emptying the center bays first, since retardant cannot backfill into the center bays from the outer bays. But if outer bays are dropped first, retardant can fill into the empty outer bays from the center bays, degrading the efficacy of the drop.[109]

DC-7 Specifications
Length: 108 ft, 11 in
Span: 117 ft, 6 in
Powerplant: 4X R-3350 radial engines
Empty Weight: 66,306 lbs
Gross Weight: 122,200 lbs
Cruising Speed: 235 knots
(359 mph in airline operations)

DC-7 Tanker 33 taxies past B-17 Tanker 89 during operations at Chico, California, in 1984. (Donald Keller)

Fairchild C-119 Flying Boxcar

The Fairchild C-119 Flying Boxcar was a natural extension of the older C-82 Packet twin-boom Air Force cargo aircraft of the 1940s. At least one C-82 is said to have flown as an air tanker; the improved C-119 made a viable air tanker in the 1970s after years of military service. A boost in performance was achieved by incorporating a J34 jet engine in a jet pack devised by Steward-Davis Company, and mounted atop the main fuselage on a number of firefighting C-119s.

C-119s tanked as much as 2,400 gallons of retardant with no visible ventral protrusions. Air tanker operators using C-119s included Hawkins and Powers, Hemet Valley Flying Service, and Aero Union. Known for its innovative air tanker conversions, Aero Union modified a number of C-119s to air tanker configuration.[110]

By the late 1980s and early 1990s, C-119s were being retired from tanker duty, with a number entering air museum collections.

Flat-bottomed C-119 transport used an internal retardant tank that exited through the structure without an external protrusion. Vents atop the boxy tank facilitate quick, clean release of retardant. (USFS)

C-119 (Military)
Length: 86 ft, 6 in
Span: 109 ft, 3 in
Powerplant: 2X R-3350 or R-4360
Empty Weight: 39,800 lbs
Gross Weight: 74,000 lbs
Cruising Speed: 162 mph

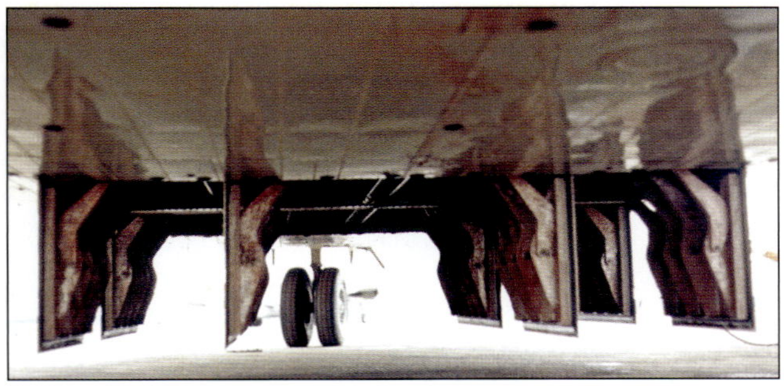

Unusual among converted transport air tankers, the Fairchild C-119 used drop gates flush with the aircraft's flat belly, with the retardant tanks carried internally. (USFS)

Some Fairchild C-119s, such as Tanker 86 (N13744), used a single J34 jet pack atop the wing to boost performance. C-119 retardant tanks were housed inside fuselage contours, with drop gates in the aircraft's flat underbelly. Hemet Valley Flying Service of Hemet, California, operated Tanker 86 when photographed at Stockton, California, in June 1978. (Author Photo)

C-119 Tanker 135 (N48076) paused at Lancaster, California, in late June 1977. (Malcolm Gougon Collection)

C-119 Tanker 81 shows areas of eddies and ablation that modify the crust of dried retardant clinging to the flat belly of the transport as it takes off on another sortie in September 1980. (Malcolm Gougon Collection)

Open inlet doors on the pod-mounted jet engine signal it is contributing its output to the takeoff power of C-119 Tanker 81 (N13743) from Fox Field, Lancaster, California, in late September 1980. (Malcolm Gougon Collection)

Fairchild C-123 Provider

Fairchild's production of the Chase-designed C-123 Provider yielded a durable twin-engine airlifter whose greatest fame came during in-country sorties in Vietnam. When modified with a pair of underwing J85 jet engines in pods, Providers were designated C-123K. The ongoing search for suitable air tanker airframes included conversion of some C-123s for the role, with tanks mounted inside the round-topped, flat-bottomed fuselage. C-123s were not major players in the air tanker fleet.

C-123 (Military)
Length: 75 ft, 9 in
Span: 110 ft, 0 in
Powerplant: 2X R-2800
Empty Weight: 29,900 lbs
Gross Weight: 60,000 lbs
Cruising Speed: 205 mph

Rarity on the firefighting circuit was this red-tailed Fairchild C-123 Provider, registered N3142D (Tanker 63), circa 1990. Gray belly paint extends well up the sides of aft fuselage, providing additional corrosion control for this retardant-dropping aircraft. (Author Photo)

Ford Trimotor

Ford 4-AT-E (Standard)
Length: 49 ft, 10 in
Span: 74 ft
Powerplant: 3X Wright J-6
Empty Weight: 6,500 lbs
Gross Weight: More than 10,000 lbs
Cruising Speed: About 100 mph

Fascinating stories about smokejumpers in Montana highlighted vintage Ford Trimotors in the 1940s and 1950s as the picturesque drop planes for the jumpers. Less well known is the conversion of two Fords in 1957 to become air tankers. Both 4-AT-E variants, Trimotors N8407 and N9612, were tanked with different systems. N8407, from Boise, Idaho, had previously been set up for aerial spraying. It became an air tanker with the addition of free-swinging doors on the bottom of the spray tanks. With two 275-gallon tanks fitted in tandem, the heavy borate load usually limited the haul to a total of 500 gallons—more than 5,000 pounds. Dropping at 50 feet, this Ford could achieve a swath about 70 feet wide and 700 feet long. Trimotor N9612, operating out of Missoula, Montana, was fitted with three 100-gallon fluid tanks. Both Fords fought wildfires in 1957.[111]

The Tin Goose, Ford's famed Trimotor, was modified with internal tanks for use as an early air tanker in the 1950s. (USFS)

Grumman F7F Tigercat

At first glance, the racy twin-engine F7F Tigercat fighter seems highly unlikely to become a successful air tanker, yet about a dozen Tigercats did so. First flown in November 1943, the Tigercat was en route to the Pacific when the war ended two years later. U.S. Marine Corps Tigercats saw combat in Korea. With a top speed well over 400 mph, good maneuverability, and tricycle-gear ground-handling ease, a long-legged F7F-3 was tested in 1958 with two modified drop tanks on wing pylons delivering a total of 1,000 gallons of retardant. Kreitzberg Aviation of Salem, Oregon, made the pioneering twin-tank Tigercat conversion. Early twin-tank configurations tended to yield less precise drop patterns, and the ultimate Tigercat air tanker incorporated a bulging 800-gallon tank attached to the fighter's belly. Surplus F7F-3 Tigercats that survived the rigors of firefighting became sleek warbirds at air shows with reversion to their original tankless military silhouettes.

With its bulbous external retardant tank this Grumman F7F Tigercat fighter looks fast even when parked. (USFS)

F7F-3 (Military)
Length: 45 ft, 4 in
Span: 51 ft, 6 in
Powerplant: 2X 2,100-hp Pratt & Whitney R-2800-34W
Empty Weight: 16,270 lbs
Gross Weight: 25,720 lbs
Cruising Speed: 222 mph

The pilot of Tigercat Tanker E23 motioned with a canteen in his hand as he waited in the cockpit for his aircraft to be reloaded for continuing fire sorties in August 1967 at Ukiah, California. In the intense summer heat, the canteen provided some refreshment for the pilot between quick-turn flights. (Author Photo)

Tigercat Tanker E63 pauses on a hosed-off ramp at Goleta, California, outside Santa Barbara in August 1967. From an early F7F Tigercat conversion using two underwing pylon-mounted retardant tanks, the F7F air tanker configuration evolved into a large bulbous fuselage-mounted tank, faired to afford some streamlining for the fast fighter. (Author Photo)

Grumman AF-2 Guardian

The culmination of Grumman's bid to succeed its own earlier Avenger torpedo bomber, the AF-2S Guardian evolved into the submarine killer portion of specialized postwar Navy hunter-killer pairs of Guardians. The hunter was the bulbous-bellied AF-2W, sporting downward scanning radar. The internal weapons bay of the AF-2S variant lent itself to conversion as an air tanker, and at least two received tanker numbers in California. E30, registered N3144G, and E21, registered N99957, were part of the growing Aero Union stable in the 1960s. By 1974, one remained flyable as an air show performer reverted to Navy blue. Several other surplus Guardians reached California, but slumbered untanked for many years at Chico.

Grumman Guardian air tanker retained the rearward-folding wings characteristic of wartime Grumman fighters and torpedo bombers, and reflected general postwar aerodynamic and structural clean-up of design concepts. Guardians saw limited use as air tankers. (USFS)

AF-2S (Military)
Length: 43 ft, 4 in
Span: 60 ft, 8 in
Powerplant: 1X 2,400-hp Pratt & Whitney R-2800-48W
Empty Weight: 14,580 lbs
Gross Weight: 25,500 lbs
Cruising Speed: 166 mph

North American AJ Savage

Early in 1960, three North American AJ-1 Savage aircraft were sold to a company that became AJ Tankers, Inc. The AJ-1 Savage was an unusual stop-gap Navy attack bomber of the 1950s that was designed to use one Allison J33 engine in the aft fuselage to augment the two R-2800 piston engines on the wings. The civilian air tanker AJ-1s deleted the jet engine. The loss of one of the civilian AJ-1s before it could see fire duty left a pair of the aircraft, registered N9142Z and N9143Z, ready to fight fires in California. Tanking 2,000 gallons of retardant, the AJ-1s were fast and large for their day, and stood out on fire duty. Initially operating out of Van Nuys Airport in Southern California, the Savages flew in the 1961 season, keeping a large front on the Bel Air fire from spreading until ground crews could reach the scene.[112]

Aircraft N9143Z, Tanker 77, fatally crashed on takeoff from Burbank, California, in September 1967. Tanker 88 (N9142Z) flew a few more times, but was scrapped around 1969. Two AJ-2s were registered N100Z and N135Z with the intent of using them as air tankers in Washington State in the early 1960s, but fire service for these two Savages did not come about. A third AJ-2, originally registered N101Z, is said to have been tanked for firefighting, perhaps in Oregon. It subsequently became a flying engine testbed for Avco-Lycoming in Stratford, Connecticut, under registration N68667.[113]

AJ-1 (Military)
Length: 63 ft
Span: 75 ft, 2 in
Powerplant: 2X 2,400-hp Pratt & Whitney R-2800-44W; 1X 4,600-lbf Allison J33
Empty Weight: 30,776 lbs
Gross Weight: 52,862 lbs
Cruising Speed: 270 mph

This AJ-2 Savage, seen in later years as a Lycoming jet engine testbed airframe, is said to have been tanked for firefighting in the 1960s. (Lycoming)

Lockheed PV-2 Harpoon

The twin-engine, twin-tail, tailwheeled PV-2 Harpoon was Lockheed's final iteration of its long-running twin engine bomber and transport line dating back to the original Electra of the 1930s. Surviving World War II and performing Naval reserve service into the 1950s, Harpoons were available for tanking by civilian operators when they were needed. PV-2s show up on air tanker rosters as late as 1983, although their numbers were never very large in the tanker fleet.[114]

PV-2 Harpoons tanked nominally 1,000 gallons. They fell out of the active air tanker fleet, as aircraft with larger capacities were available to replace them.

Paddle props capped with non-military spinners pulled this PV-2 to fires when the tanker was photographed at Wenatchee, Washington, in November 1972. (Author Photo)

PV-2 (Military)
Length: 52 ft, 0.5 in
Span: 74 ft, 11 in
Powerplant: 2X 2,000-hp Pratt & Whitney R-2800-31
Empty Weight: 21,028 lbs
Gross Weight: 36,000 lbs
Cruising Speed: 171 mph

PV-2 Harpoon Tanker 29 (N83L) rests at ease, its drop gates open. Photo was probably taken at Medford, Oregon, in the 1960s or 1970s. (Kenn Knackstedt photo via Donald Keller)

PV-2 Harpoon Tanker 43 extended the line of its retardant tank lower than the original bomb bay line, as photographed in Wenatchee, Washington, on 23 September 1972. Retardant supply hose in foreground rests on wheeled dollies to facilitate quick movement on the tarmac. (Author Photo)

Tanker 40, a low-slung PV-2 Harpoon dating back to World War II, settled in at Wenatchee, Washington, after a sortie in July 1970. (Author Photo)

Lockheed PV-2 Harpoon, last in an evolution of twin-engine, tailwheel patrol bombers the company produced during World War II, served in limited numbers as an air tanker. This example, Tanker 40 (N6651D), was part of the Wenairco operation in Wenatchee, Washington, when photographed in August 1971. (Author Photo)

Simple angular sheetmetal fairings wrap the lowered front of a PV-2 Harpoon retardant tank. Removal of Harpoon's original bomb bay doors to accommodate the retardant tank resulted in a curved gap above the tank. (USFS)

Lockheed P-2 Neptune

The Lockheed Neptune patrol bomber straddled the change in U.S. Navy aircraft nomenclature, entering service as the P2V and leaving as the P-2, with suffix numbers or letters denoting model. Early efforts at using P-2s in the United States, circa 1970, involved surplus Neptunes leased by the Forest Service to air tanker contractors. Soon, tanker companies were buying surplus Canadian Neptunes outright.

By the late 1960s, the USFS had steered away from using surplus single-engine air tankers like the TBM (although other fire protection agencies continued to use single-engine air tankers well into the 1970s). Interest was high in determining if newer, faster, and more capacious multi-engine aircraft would increase the effectiveness of the air tanker mission.[115]

A likely candidate was the P-2 Neptune, a durable patrol bomber with an internal weapons bay, and a mixed propulsion system of two R-3350 piston engines supplemented by two underslung wing-mounted J34 turbojets, which burned aviation gas like the piston engines. The jets could be used in phases of flight where extra power and acceleration were required, and could be shut down when their extra thrust was not needed. First evaluation by the Forest Service of a surplus Navy P2V-5 (EP-2E) Neptune fitted as an air tanker took place at Medford, Oregon, in October 1969. On 24 November of that year, the Neptune had its first actual firefighting opportunity when a late season wildfire presented itself. According to a contemporary document: "Reports from ground personnel, two air tanker bosses, and a NASA Velocity Gravity Height (VGH) recorder confirmed the belief of project personnel and Rosenbalm Aviation that this aircraft is the most promising air tanker evaluated to date."[116]

The Forest Service obtained three P-2 Neptunes from Navy storage at Davis-Monthan Air Force Base in Tucson, Arizona. Arrangements were made with Rosenbalm Aviation of Medford, Oregon, to take the three Neptunes out of their preserved storage condition and ferry them to Medford. When one of the Neptunes proved to have mechanical problems, the other two were made ready and flown to Medford. The Forest Service gave Air Tanker Contract No. 002665 to Rosenbalm Aviation, a seasoned air tanker operator, to conduct a multi-part evaluation and modification of one Neptune, registered as N126Z.

First flight operations were test flights of the unmodified Neptune to simulate air tanker maneuvers in the mountainous region surrounding Medford. Still essentially a military patrol aircraft, the first flights impressed air tanker pilots because of the Neptune's spoiler-aided maneuverability, a plus for mountain flying. Documentation from this period included the observation: "The aircraft handled very well in all flight regimes during this test. Simulated drops were made in canyon bottoms at 120 knots after which power was applied to all four engines. The aircraft accelerated very rapidly, and climbing out of canyons was not difficult."[117]

P-2H (P2V-7) (Military)
Length: 91 ft, 4 in
Span: 103 ft, 10 in
Powerplant: 2X 3,500-hp R-3350-32W;
2X 3,400-lbf J34-WE-34
Empty Weight: 47,456 lbs
Gross Weight: 75,500 lbs
Cruising Speed: 207 mph

Dark-bellied old-style Navy paint scheme covers a surplus P2V-5 Neptune ferried to Medford, Oregon, for tanker conversion under the original Forest Service plan. Ultimately, it may have been scrapped as a different means of acquiring Neptunes, including foreign sales, which opened the market to private buyers. (Carl M. Johnsen)

Early test drop by the first Neptune air tanker, the Forest Service's N126Z, was made while the patrol bomber still retained most of its U.S. Navy markings. (Kenn Knackstedt)

Still wearing some of its Navy markings, the prototype P2V-5 Neptune air tanker makes a demonstration drop very close to other surplus Neptunes and an A-26 at Medford, Oregon, in 1969. (Kenn Knackstedt photo via Donald Keller)

A flight test report filed 15 October 1969 quantified typical engine management expected for the P-2 air tanker: "Fuel consumption per working hour is 400 gallons. Jets are used at 100 percent until landing gear is up, then 94 percent to cruise altitude at which time they are placed in idle to the fire. On the fire, they will be used as needed; leaving the fire the jets will be shut down." Radial engines consume oil in operation. The Neptune tanked 60 gallons of oil for each Wright R-3350 engine, and oil consumption was figured at about 1 gallon per hour.[118]

The early evaluation flights were followed by the installation of a retardant tank in the Neptune's weapons bay. In about four days, the Rosenbalm team stripped more than a ton and a half of unneeded military gear from the Neptune. Working with Lockheed's Joe Paul, Rosenbalm Aviation built and installed a 3,000-gallon tank in the Neptune's fuselage bay. Six individual 500-gallon tanks comprised the installation. Doors were hydraulically operated by two redundant systems independent from the aircraft's hydraulic system. Six drop switches, one for each door, were mounted on the pilot's control wheel in this pioneer Neptune. About 2,700 man-hours were expended from the time the Neptune was picked up at Arizona until it was ready to flight test as an air tanker.[119]

Next, drop pattern suitability was evaluated. After trying measurements of water in a grid of cups placed on stakes, the team decided to drop on the ramp and taxiway area at Medford. "This permitted visual observation, timing, and measurement, and seemed to work out very well," a paper on the project noted. Various drop patterns were achieved at a drop height of 100 feet and a speed of 120 knots. "In a train drop of two 1,000-gallon drops of water at the same speed and altitude, the

P-2 Neptune retardant tank shows multiple gates open. Fairings fore and aft provide a measure of streamlining. (USFS)

Carrying bold USFS lettering at the base of the vertical fin, the first air tanker Neptune was this P2V-5 (N126Z). It participated in the central Washington fire battles of July 1970, and flew from Wenatchee, Washington, during that time. (Author Photo)

area covered with standing water measures 600 feet long and 60 feet wide." Through a series of 500-gallon, 1,000-gallon, and 3,000-gallon drops, the Neptune proved to handle very well while releasing liquid. Unlike some air tankers, with the pioneer Neptune "there is no tendency to pitch up after the drop," a report noted.[120]

On a temporary basis, the Neptune was to be used on real fires when possible during the 1969 season. The opportunity came that November. Spanning an hour and 15 minutes flight time, the first Neptune air tanker made three water drops with its first tank load, and three retardant drops in train with its second tank load. Observers liked what they saw, and were impressed with the P-2's cross-country speed to the fire, its maneuverability, and the large amount of liquid available to release. A NASA recording device on the Neptune subsequently verified all flight loads were well within the aircraft's normal flight envelope; maximum g-load recorded was 1.8.[121]

The pioneer Neptune air tanker was registered as a Public Aircraft operated by the USFS under U.S. civil registration number N126Z. Its Navy Bureau Number was 124904. Enthusiasm was high for using Neptunes as air tankers; some pointed out a single Neptune could carry as much retardant as three twin-engine A-26 Invaders. There was a potential downside to the introduction of an air tanker as capable and heavy as the P-2: some of the smaller air tanker bases had ground weight limitations that would reduce the P-2's allowable tank load to something less than the maximum 3,000 gallons. One solution offered was to base the faster Neptunes at more centrally located airfields that could handle their weight.[122]

Early P-2 air tanker pilot Lou Remschner summed the Neptune's advantages in a 1971 letter: "…it is difficult to compare the P2V with any other tanker. I think to make an actual comparison would be one P2V compared to three B-26s or four F-7s (F7F Tigercats) or one and a half PB4Ys or a fleet of TBMs for getting retardant to the fire." Remschner noted: "…the most comfortable feeling is the fresh equipment, late model engines, airframes, etc., on the P2V. All emergency equipment is easy to operate, the engine warning systems give you ample time for corrective action, and all normal systems are simple and easy to operate. The P2V is a comfortable aircraft to fly and work with its low noise level."[123]

Retired Air Force test pilot Lt. Col. Richard Bradley evaluated his brief stint in the right seat of a B-17 air tanker and the P2V air tanker: "The differences over a fire during attack are immediately apparent. The B-17 has a slow response in roll rate with accompanying high aileron control force; the same applies in pitch control at drop and slow acceleration after drop…. The incorporation of spoilers on the P2V provides a rapid roll rate response, which is very effective in applying last moment lateral corrections prior to drop. Also in counteracting violent heat thermals… The rapid response of the J34s, in addition to the R-3350s, provide the acceleration desired to overcome an engine failure after drop which in some instances would be marginal in a B-17."[124]

If the B-17 was a mighty player in the development of air attack capability, and a sentimental favorite of those who watched air tanker operations as enthusiasts, the advantage of newer, larger air tankers like the Neptune was the inevitable harbinger of the next generation.

Colorful Neptunes at Minden, Nevada, on 27 April 2001 show several solutions to rounding off the aft fuselage. Early Neptune patrol bombers had a gun turret in the extreme aft fuselage. P2V-5 and P2V-7 developments typically replaced the guns with a long boom housing Magnetic Anomaly Detector (MAD) gear for locating submarines. Bright yellow Tanker 48 in the background retains original MAD boom contours; Tanker 99 in the foreground features a truncated tail stinger for the aircraft's role as an air tanker. The prototype USFS Neptune conversion (see photos on page 60) flew with the MAD boom removed; its fuselage terminated in a flat plate. (Author Photo)

By the end of the 1973 fire season, Forest Service Region 1 in Missoula, Montana, had two years' experience working with another converted Neptune, a surplus Canadian Armed Forces P2V-7 (SP-2H) registered to Johnson Flying Service of Missoula. Canadian surplus Neptunes became available for purchase directly by air tanker companies. This early repatriate carried U.S. civil registration N14448, formerly Canadian civil registration CF-MQW. Twenty-five military P2V-7 Neptune derivatives were ordered by Canada in 1953, initially intended to be operated without the underwing jets. Fortunately, these aircraft received the jet engines that proved so advantageous in firefighting operations years later.

The Federal Aviation Administration (FAA) certificated N14448 in the Restricted category "to perform the special purpose operations involving Forest & Wild Life Conservation (Aerial Fire Fighting)." In this category, the aircraft was prohibited from carrying people or cargo for hire. Only the minimum required aircrew could be aboard during actual fire operations. Maximum takeoff weight was not to exceed 80,000 pounds, "and shall be adjusted downward if necessary, to provide a safe margin of performance under existing operating conditions, considering the takeoff area, altitude, temperature, and terrain, so that the operation will not endanger life or property."[125]

The Johnson Flying Service modifications to this SP-2H were listed as: "Removal of all military equipment such as electronic gear, external radomes, external store mounts, wingtip tanks, Magnetic Anomaly Detection (MAD) tail boom, and the installation of a 3,000-gallon retardant tank in the bomb bay."[126] The Neptune's legendary internal fuel capacity of 2,800 gallons made the tip tanks superfluous to air tanker operations. The long tapered MAD boom stinger was sometimes capped with a blunt plate when removed from air tanker Neptunes; some latter-day conversions featured a shorter streamlined stinger, or a rounded end cap.

The Forest Service's Wildland Fire Chemical Systems (WFCS) program evaluated a late-model P2V-7 Neptune to quantify its retardant ground pattern. The WFCS report described the test Neptune, from Neptune Aviation: "The conventional tank is

Tanker 18, a P2V-7 derivative, has the high, bulged canopy that helps define the Dash-7 version of the Neptune. (Author Photo)

Tanker 12 shows its wingtips, cleaned of Navy tip-tanks, as the Neptune departs Fox Field in May 1994. (Author Photo)

constructed of aluminum. The door opening is actuated by an independent hydraulic system mechanically activated using 28 volts DC aircraft power using a Black Hills-type controller. The tank is certified to hold 2,450 gallons. It includes six compartments and six doors arranged side-by-side, running the length of the tank with all doors hinged on the outboard side. Each compartment contains deployable flow restrictors to help control the flow rate. Restrictors serve as a physical barrier that lowers the flow rate as liquid is released from the tank."[127]

Science adds to the precision effectiveness of modern air tankers as physical barriers can be employed to optimize the drop for specific fire conditions. For the Neptune tests, the Missoula, Montana, Technology and Development Center placed an array of plastic bowls at a test drop site, measuring the quantity of liquid in each container as evidence of drop pattern effectiveness. The testers explain: "Flow rate, drop height, and airspeed affect the drop pattern. Increasing drop height gradually widens the drop while decreasing the coverage levels. Airspeed has a much greater effect on the drop pattern. Increased airspeed increases line length while reducing the coverage level. Because each compartment has restrictors, specific coverage levels can be produced by raising or lowering restrictors and releasing more than one compartment at a time."[128]

Fire managers have tools including fuel fire behavior predictors that can adjust the amount of retardant appropriate for each 100 square feet of terrain to be covered. Annual and perennial western grasslands and tundra typically need 1 gallon per 100 square feet; forest slash burning can demand more than 6 gallons per 100 square feet. Depending on the need, the P2V-7 in the tests could deliver 1/2 gallon of coverage per 100 square feet in a line 1,740 feet long, or it could saturate an area with 8 gallons per 100 square feet for a brief 40-foot line. These figures were based on unrestricted flow, a drop height of 171 feet, an airspeed of 122 knots, and a wind speed of 14.5 mph. Various time intervals between tank releases were used to get the desired coverage. When drop height was raised to 196 feet, flow restrictors were deployed, and time intervals between release were adjusted, the 8 gallon coverage could be sustained over a line 220 feet long. Both sets of tests used 808 gallons of gum-thickened retardant, roughly one third of the aircraft's capacity.[129]

P-2s survived the 2002 stand-down of air tankers and remained a strong part of the air tanker fleet in subsequent fire seasons.

Tanker 11, N14447, shows another version of a bobbed P-2 tail boom; several options have been tried on elements of the Neptune fleet. (USFS)

Lockheed P-3 Orion

Aero Union of Chico, California, has developed tanks and systems to make the Lockheed P-3 Orion patrol bomber an effective large air tanker. In Canada, Air Spray Ltd. has modified similar commercial Lockheed L-188 Electra airframes with Aero Union tanks.

The P-3/L-188 can hold 3,000 gallons of retardant. The Aero Union tank has a constant-flow capability that distributes retardant uniformly for more effective fire coverage. The large tank is a single compartment with bulkheads that serve dual purposes of diminishing sloshing fore and aft while providing structural integrity. Two doors, opening from the center, control flow rate by how far they are opened. An automatic computerized controller senses the level of retardant in the tank and constantly adjusts the door opening to maintain the desired flow rate—a far cry from simple gravity drops in the early years of firefighting. The pilot of the P-3 or L-188 with this computerized tank can select any of eight flow rate settings to match the requirements of different wildfires.[130]

As an example of this aircraft's capabilities, the Forest Service used accepted models for fire fuel types to predict that a fire in a short needle conifer forest with heavy dead litter on the ground could best be treated with the selector indicating a flow rate of 698 gallons per second. At this rate, a drop of 1,500 gallons would produce a swath 349 feet in length with little variation in uniformity.[131]

Like their older P-2 ancestors, P-3s emerged as strong components of the air tanker fleet following reviews of the air tanker mission in 2002 when two other types of large air tanker were lost to inflight break-ups.

P-3 Tanker 00 shows the patterns made by colorful retardant on its lower surfaces from releases as well as venting. (Author Photo)

Crisp P-3 Orion N900AU (Tanker 00) whistles in for a landing at Lancaster, California, circa 1996. Aero Union has made the P-3 a viable heavy air tanker. (Author Photo)

Three thousand gallons of retardant make the P-3 Orion, with its computer-controlled drop gates, a potent air tanker capable of laying a long, steady fire line to help ground firefighters encircle a blaze. (USFS)

P-3C (Military)
Length: 116 ft, 10 in
Span: 99 ft, 8 in
Powerplant: 4X 4,910-ehp Allison T56-A-14
Empty Weight: 61,491 lbs
Gross Weight: 142,000-lb overload
Cruising Speed: 237-mph patrol speed

Lockheed C-130 Hercules

The workhorse Hercules turboprop transport served as a civilian-operated air tanker with belly discharge doors, and as a convertible Air National Guard version with the temporary installation of the Modular Airborne Firefighting System (MAFFS). First-generation MAFFS versions discharged retardant off the back ramp of the C-130; newer variants use discharge pipes out the paratroop jump door in the side of the aft fuselage. The Hercules is rated as a 3,000-gallon air tanker.[132] Early C-130A models converted by civilian air tanker operators were withdrawn from federal contracts in 2002 following the inflight break-up of a C-130 on a fire in California.

Starting in 1971, Air National Guard and Air Force Reserve C-130s have participated in annual wildfire suppression missions with civilian agencies. Often, the military C-130s are called only "when all commercial equipment has been used or there is a threat to life," according to the Air Force. "The U.S. Forest Service then arranges for Air National Guard/Air Force Reserve fire fighting assistance. Personnel and equipment are then under the direction of the Forest Service."[133]

Early generation MAFFS equipment used five pressurized retardant tanks and "two large exhaust manifolds for spraying a dual purpose fire retardant compound from the rear of the aircraft. The compound is also a fertilizer; it stifles flames and later stimulates plant growth. It takes 20 minutes to fill the tanks with chemical. Release of the entire load over the fire area takes about eight seconds."

MAFFS was originally designed by the FMC Corporation's Defense Technology Laboratory in San Jose, California, as part of an Air Force contract. The system was developed and built in rapid fashion between March and October 1971, when MAFFS had its first use on an actual fire.[134] From a test installation using nozzles connected to 500-gallon tanks, the production version was initially slated to carry 2,500 gallons, subsequently increased to 3,000.[135]

MAFFS went overseas with the Wyoming Air National Guard in October 1997 to render assistance to forest firefighters in Indonesia. Australia, Greece, Portugal, and Italy have all used MAFFS systems to fight wildfires.[136]

Aero Union picked up the MAFFS system and developed it further. By July 2007, Aero Union finished ground testing the next generation MAFFS II airborne firefighting system. Suitable for C-130 E, H, and new J-models, MAFFS II can carry 3,400 gallons of retardant in a pressurized single tank. The retardant exits through a nozzle in a special paratroop door plug at the side of the fuselage to the rear. This plug allows the

C-130E (Military)
Length: 97 ft, 9 in
Span: 132 ft, 7 in
Powerplant: 4X T56-A-7A turboprops
Empty Weight: 72,892 lbs
Gross Weight: 175,000 lbs
Cruising Speed: 340 mph

Five tanks and two pipes of a MAFFS system roll aboard an Air National Guard C-130 seasonal air tanker in 1985. (Donald Keller)

C-130 to remain pressurized "for the comfort and safety of the crew during aerial firefighting missions," according to Aero Union's Robert Farinsky.[137]

The MAFFS II system can be rolled on and off the C-130 with no time-consuming modifications, enabling quick conversion to and from the fire suppression mission given to the Air National Guard. MAFFS II can deliver retardant in variable amounts, depending on the nature of the fire. This includes a saturation of 8 gallons of retardant for every 100 square feet of surface to be covered, doubling the coverage possible with older MAFFS units.[138]

MAFFS system deluges the ramp at Chico, California, in this 1985 test. (Donald Keller)

Hastily applied DayGlo numbers and stripes identify an Air National Guard MAFFS C-130 Hercules. (USFS)

Photo shows a development of the MAFFS retardant system using two pipes dumping downward off the ramp of a C-130. Roll-on, roll-off MAFFS retardant tanks can be seen dimly in the fuselage. (USFS)

C-130A (N133HP, Tanker 133) with small underwing external fuel tanks featured a drop system different from the removable MAFFS concept. Hawkins and Powers Aviation operated this clean air tanker. (USFS)

MAFFS C-130 test drop shows retardant spewing from pipes mounted in the side troop doors. Iterations of MAFFS have been attractive augmentations to the annual American air tanker force, permitting military C-130 transports to become seasonal air tankers. (USAF)

Boeing C-97 Stratocruiser

Firefighters displayed an on-again, off-again relationship with the giant double-deck Boeing C-97. An early YC-97, with R-3350 engines and a smaller vertical tail comparable to that of a B-29, was tanked to carry 4,500 gallons, and operated against a few fires in the 1961 season. The aircraft's performance was acceptable, but it is said some fire officials were reluctant to call in such an expensive piece of equipment.[139]

In the mid-1980s, the availability of later C-97 variants with R-4360 engines and the taller vertical fin associated with production C-97s prompted a renewed interest in the type as an air tanker. Hawkins and Powers Aviation operated C-97G Tanker 84, later as Tanker 97, in Alaska for a number of years. Possibly the impending availability of surplus C-130 Hercules turboprop transports eclipsed interest in C-97s; the C-97 was destined to remain a rarity at air tanker bases.

C-97 (Military)
Length: 110 ft, 4 in
Span: 141 ft, 3 in
Powerplant: 4X R-4360-59
Empty Weight: 82,500 lbs
Gross Weight: 175,000 lbs
Cruising Speed: 300 mph

Rare photos of the Boeing YC-97, one of four prototypes built of the rugged C-97/KC-97 Air Force cargo transport and tanker series. The YC-97 was unique in having the B-29's Pratt & Whitney R-3350 engines and shorter vertical stabilizer rather than the C-97's more powerful P&W R-4360s and taller tail from the B-50. YC-97s were granted FAA Restricted Type Certificates in 1957, and were operated for a brief time as fire bombers along with the B-25, Privateer, and TBM shown in the top photo. Oddly enough, of these four aircraft, only the Privateer continued being used as a popular air tanker to the end of the twentieth century. (USFS via Author's Collection)

Grumman S-2 Tracker

Conversions of piston-engine Grumman S-2 Tracker patrol planes yielded air tankers with a retardant load of 800 gallons. The State of California Department of Forestry has been a major user of S-2 Trackers for decades. Marsh Aviation delivered an upgraded version to California that uses turboprop engines, a more pointed streamlined nose, a new electrical system, new avionics, and a larger 1,200-gallon constant flow retardant tank system. With faster cruise speeds and higher capacity, these Turbine Trackers can deliver more retardant to a fire in a given time period than can the older piston-engine Trackers.[140]

The State of California obtained 19 surplus S-2 Trackers from the Department of Defense (DOD) in 1972, following an evaluation by the Army's aviation test operation in California that declared the Tracker a suitable replacement for the venerable TBM Avenger. The state had some concerns about the continued viability of the aging TBM fleet. The prototype S-2 air tanker retardant tank system was built at the California Department of Forestry (CDF) Mobile Equipment Facility at Davis, California. Hemet Valley Flying Service, an experienced air tanker operator, modified the first two designated S-2As to enter service as air tankers in California. Some cross-border assistance for the design of the modification came from Ontario Lands and Forests in Canada. The first of California's Trackers flew on fires in 1973, and became a familiar sight in the skies over California fires for decades. A state history of air tankers says: "Three TBM accidents in 1973 and three F7F accidents in 1974 accelerated the CDF S-2 modification program. As a result, contracts were entered into with four California contractors, Aero Union Corporation, Sis-Q Flying Service, TBM, Inc., and Hemet Valley Flying Service to modify and tank 10 S-2 airtankers during the 1973/1974 winter period." A dozen S-2As were in service over California by 1974, with five more built up by Bay Aviation Services in time for the 1975 fire season. The Trackers were obtained on three lease agreements between California and the U.S. Navy, bringing a total of 55 S-2s and 60 engines into the state program. The Tracker tanker was to be operated by a single pilot.[141]

In 1987, CDF contracted with Marsh Aviation in Mesa, Arizona, to update an S-2 Tracker air tanker

S-2E (S2F) (Military)
Length: 43 ft, 6 in
Span: 72 ft, 7 in
Powerplant: 2X 1,525-hp Wright R-1820-82-WA
Empty Weight: 19,033 lbs
Gross Weight: 26,867 lbs
Cruising Speed: 149 mph

Operated by Hemet Valley Flying Service for the California Department of Forestry, Tanker 71 (N403DF) is a standard piston-engine S-2 Tracker air tanker. Elevator pushes nose down in this photo as retardant load lightens the ex-Navy carrier aircraft. (USFS)

Tanker 76 (N417DF) was a California State S-2 when photographed at Stockton, California, in June 1978. Retardant tank line protrudes slightly below fuselage contours. (Author Photo)

with turboprop engines and greater load capacity. It flew operational missions from several bases. In 1996, CAL FIRE (the California Department of Forestry and Fire Protection's newer moniker) acquired 26 S-2E and G-model Trackers from DoD for conversion by Marsh Aviation into so-called S-2T turbine variants. The last of the batch were delivered in 2005. Two Garrett TPE331-14GR turbine engines in streamlined composite cowlings provide 1,650 hp each, and are a visual cue to identify turbine Trackers. Noses on California's S-2Ts have been streamlined and extended into a point. Maximum operating speed is 270 mph, considerably faster than a piston-engine Tracker. This enhances the S-2T's delivery capability on a fire by shortening the duration of outbound and inbound flights. A range penalty is a trade-off with the turbine Trackers; piston S-2A tankers had a listed range of 800 miles loaded and 1,000 miles empty, while S-2Ts have a loaded range of only 500 miles and an empty range of 800 miles. Nonetheless, most fires are within a shorter operating range to an air tanker base. As with the piston-engine Trackers before, California uses the S-2Ts for fast initial attack.

Glowing lime green paint with orange trim was an early version of California Division of Forestry's see-and-be-seen color scheme for its S-2 Tracker air tankers. This photograph was taken in September 1980. (Malcolm Gougon Collection)

Over a field of test measuring cups, a nearly stock Grumman S-2 Tracker releases retardant for evaluation of pattern. The USFS conducts tests to determine the effectiveness and best drop speeds and altitudes for the various air tanker types. This information guides pilots and planners in how to best employ the available tankers against varying fire and terrain situations. (USFS)

Change in 2004 Federal Contract

Three inflight wing failures of air tanker aircraft—a C-130A in 1994, another C-130A in 2002, and a PB4Y-2 Privateer also in 2002—prompted questions about air tanker airworthiness. It was a difficult time in the industry—operators and aircrews worked long hours and braved many dangers to protect wildlands from fire; their defenders did not want to see them blamed for mishaps that might be the result of factors beyond the airworthiness standards they were expected to meet.

On 19 July 2002, the day after the Privateer mishap, the Aerial Firefighting Industry Association (AFIA) issued a bulletin saying: "All operators fly under strict FAA approved maintenance programs." The bulletin also anticipated the future: "There is an initiative to modernize the fleet over the next three to five years to eventually replace the older aircraft being used, primarily due to the diminishing availability of long-term spare parts." Exceptionally prolific fires were noted: "As of June 30 of this year (2002), due to the intense level of fire incidents the fleet as a whole has been flying at a rate which is double the norm in terms of flight hours."[142] In another bulletin issued that December, the AFIA cited "the need for government wildland firefighting agencies to demonstrate the cost-benefit of the large airtanker program to Congress to ensure the fire program is funded at a level that supports a world-class airtanker program..." AFIA also said there was a "need for government wildland firefighting agencies to validate... long-range aerial firefighting requirements as early as possible so that industry has the necessary time to plan, convert, and build the support structure to meet operational needs."[143]

The USFS and the Bureau of Land Management (BLM) convened a blue-ribbon panel to study the state of air tanker operations following the 2002 crashes. The panel's marching order was to identify essential information to allow planning a safe and effective future aviation program. Written and verbal comments were obtained from air tanker operators as well as representatives of involved government agencies including the FAA, NTSB, CDF, and others. Air tanker operators Aero Union and Conair briefed the panel. The panel presented eight findings, which the report's executive summary characterized as "strong inferences rather than deduced certainties." The panel's mandate "was to identify problems, not advocate solutions or make recommendations."[144]

The panel said the safety record of fixed wing and rotary wing aircraft used in wildland fire management was unacceptable. And the finding included Beech Baron BE-58P bird dog aircraft owned and operated by the USFS—not just privately operated air tankers were criticized. The panel concluded, "...the federal government is asking employees involved in aerial firefighting to take unnecessary risks. However, it expects contractor crew members to assume an even higher degree of risk than government employees." The panel acknowledged a "general, long-term improvement in accident rates" for air tankers, but said the number of large air tanker crews killed since 1958, if expressed as a percentage, would be equal to more than 200 ground firefighter deaths each year—a statistic that clearly would not be tolerated. The panel also found the nature of wildfire control had assumed importance "that warrants the attention of national leaders." The panel noted, "...controlling wildland fires cannot be considered an auxiliary mission second to land management."

The intensifying wildland-urban interface demanded different fire tactics in built-up areas than in open wildlands. More than 15 million acres of wildland burned in the United States between 2000 and 2002, almost doubling the 10-year average for burned acreage. Add a newfound concern for homeland security, and the wildfire problem took on new proportions. "The environmental conditions that created today's threat of large fires have taken decades to develop and are not the fault of any political party or federal land management agency," the panel noted with some diplomacy. "The conditions evolved from uncoordinated policy changes that responded largely to public demands and have been exacerbated by shifting demographics, severe drought, and a drastic change in the national security landscape." The panel observed there was another time when wildland fire protection took on national security emphasis:

Trading its usual garland of Douglas fir for Eucalyptus, **Hawaii Mars** ***rested on California's Lake Elsinore in October 2007.*** (Author Photo)

"During World War I, the new (fire control) policies took on a national security perspective when leaders feared enemy saboteurs could easily start hundreds of massive wildland fires, overwhelming the country's firefighting capabilities.[145]

Many years of expectations, from the public as well as other forest land users, had evolved a program of containment and extinguishing wildfires so thoroughly that natural burn-offs of undergrowth could not happen, resulting in too much fuel available for super fires that subsequently fed on the abundance of natural fuel. A 2000 report on wildfire management cited by the panel described the problem succinctly: "While the policy of aggressive fire suppression appeared to be successful, it set the stage for the intense fires that we see today. Full suppression of all wildland fires initially gave our forests and wildlands a chance to heal, creating a false sense of security. However, after many years of suppressing fires, thus disrupting normal ecological cycles, changes in the structure and makeup of forests began to occur. Species of trees that ordinarily would have been eliminated from forests by periodic, low-intensity fires began to become a dominant part of the forest canopy. Over time, these trees became susceptible to insects and disease. Standing dead and dying trees, in conjunction with other brush and downed material, began to fill the forest floor. The resulting accumulation of these materials, when dried by extended periods of drought, created the fuels that promote the type of wildfires that we have seen this year (2000)."[146]

The previous half-century saw a marked increase in the number of people moving into rural areas, intermingling into wild areas where human habitation formerly was nonexistent. This altered the ways fire managers could address wildfires; previously, some rural blazes could be allowed to burn out; now, human habitation in these areas demanded action. And the cost of assigning ground and aerial resources to protect small enclaves was high; one USFS firefighter told the blue ribbon panel: "We could have bought [one] entire town for what we spent protecting it." The point was made—people and structures raise the ante in fire suppression costs and obligations.[147]

One measure of the increase in activity is flying hours. In the 1980s, for an air tanker to log 100 flying hours in a season was remarkable, and 200 hours was very rarely seen. The next decade, air tankers averaged 250 to 400 hours each season, the panel noted, "and the

Coulson Flying Tankers brought a tank truck to service the classic* Hawaii Mars *on its 2007 duty at Lake Elsinore in California. (Author Photo)

trend toward 400-hour years has continued." Factoring into the higher flying hours per season are longer seasons fueled by drought conditions for a number of years. Long seasons and monster fires put large air tankers in the unintended business of ongoing firefighting against a blaze instead of the intended function of initial attack until ground firefighters could get the upper hand. All this has had an adverse affect on airframes, engines, and aircrews, the panel noted.[148]

The panel was succinct in its description of air tanker operations as of 2002: "Under the current system of aircraft certification, contracting, and operation, key elements of the aerial wildland firefighting fleet are unsustainable." The state of air tankers was described in the panel's report: "This is the end of a third type of aircraft cycle, which have each culminated in accidents and crew fatalities. Each cycle begins with the Forest Service and its private-sector contractors obtaining retired military bombers, transports, and patrol aircraft from the Pentagon's 'boneyard.' Stripping out unneeded systems to reduce weight, contractors outfit the air tankers with fire retardant tanks and delivery systems, obtain FAA

P-3 Orion N927AU, Tanker 27, returns from a drop mission circa 1994 with retardant stains blending in with the bright red-orange paint Aero Union employs to make air tankers visible over a fire. (Author Photo)

Restricted Category type certificates, and put the aircraft into firefighting service. There are a few nonmilitary exceptions, such as FAA-certified commercial air transports converted to air tankers, but, to the panel's knowledge, none of these have suffered catastrophic structural failures."[149]

The rigors of wildfire air drops, coupled with the legendary fortitude of air tanker pilots, was quantified to an extent in the 1970s and 1980s by studies that instrumented several large air tankers to record stresses on the airframes. Seven monitored, large air tankers posted average drop speeds faster than the maximum allowed speeds for those aircraft in a study that ran from 1983 to 1989. Exceeding 4 gs happened on average once every 100 air drops, while acceleration limits were exceeded 20 percent of the time, according to data cited from the *1983 Operational Retardant Evaluation Study,* Fire Suppression Research Report 1990.[150]

A study of P2V Neptune air tankers found one aircraft with a 13-inch wing crack, and another that experienced a 5-inch growth of a crack in the course of one day of firefighting. The blue-ribbon panel reported: "The firefighting mission, especially in mountainous terrain, forces pilots to routinely exceed g-limits that large air tanker aircraft usually would not expect to reach but a few times in the aircraft's life."[151] If studies show wing crack problems on a number of air tankers, it is well to remember that some aircraft in a variety of roles other than aerial firefighting also sustain airframe cracks at some point in their lives; if cracks are problematic in the air tanker fleet, they are not unique to that operation. During the blue-ribbon panel's information-gathering panels, some air tanker operators emphatically disputed the idea that aerial firefighting produced a flight environment more severe than carrying and delivering cargo.

Ideas that came out of the panel's efforts included the instrumenting of one of each type of large air tanker to act as a weather vane for predicting fleet issues of airframe degradation. A company engaged in work of this type suggested it could cost around a half-million dollars for each instrumented air tanker, a cost that would have to be factored into the overall wildfire budget picture.

"Private operators, for the most part, have done an admirable job of keeping these aging aircraft flying," the panel noted. "However, they are handicapped by receiving little, if any, support from former military operators and the aircraft's original manufacturer. Additionally, the FAA has essentially said, 'It's a public-use aircraft. You're on your own.' There are few checks and balances to ensure that the aircraft are airworthy and safe to fly throughout a fire season. Contractors have no financial incentive and are not required to ensure that their aircraft are safe to fly."[152] Nonetheless, defenders of air tanker operators would say they did keep their aircraft safe within the guidance available to them at the time.

The panel decried "mission muddle" in which the variety of fire missions, philosophies, and unclear standards of federal land management agencies "seriously compromises the safety and effectiveness of aviation in wildland fire management." The panel reported its belief "that aerial firefighting tasks remain higher than necessary because the mission differences among agencies have not been recognized, reconciled, and

In about 1994, Hawkins and Powers' C-130 Tanker 133 (N133HP) returned to Fox Field in California, its belly stained by retardant issued from its flush-mounted gates. (Author Photo)

expressed as a common operations plan with clear lines of authority... Jurisdictions and federal- and state-agency responsibilities must be reconciled to develop a more effective national wildland aerial firefighting program." Some agencies were already engaging in interagency coordination and cooperation, but at other times, bureaucracy got in the way of decisiveness. "...it was obvious to the panel that there is no clear, general understanding of the degree to which fire detection and initial attack are priorities. A strong consensus of those talking to the panel was the fear that "a reduced emphasis on initial attack is harmful to the firefighting mission of land management agencies."[153]

One Forest Service official was quoted by the panel as saying "people don't think you're really fighting a big fire," unless they see air tankers in the news. The panel heard: "Large air tankers are sometimes used where they are not particularly effective because people, news media, and public officials equate fire bombers dropping retardant with aggressive firefighting." To the extent this is done, it detracts from the sound use of air tankers for initial attack, the panel concluded. California's state aerial firefighting program was cited as an example of proper initial-attack philosophy, followed by scaling back air assets as ground crews took over.[154]

Federal agencies tasked with wildfire management came in for some criticism from frontline aerial firefighters over multi-agency committees that must achieve consensus during fire operations. "The feds don't fight fire in the morning because all the leaders are in meetings," the panel was told.

The panel's report opined: "The apparent, but possibly illusory, reality that funding is never sufficient has bred a culture that accommodates risk in aerial firefighting activities. For the aviation program overall, this has translated to insufficient contract funding to provide adequate knowledge of aircraft condition; insufficient training, inspection, and maintenance; and a deplorable safety record for large air tankers. *A culture that emphasizes cost-efficiency has also created an admirable, but hazardous, 'can-do' ethos that pervades firefighting aviation.*" (Emphasis added.) The panel observed: "Unwittingly, the Forest Service has exploited the passion and willingness of its firefighters to do more with less."

It seemed as if the very spirit of independence and ingenuity that gave birth to air tankers was being blamed in part for the problems the panel observed. The magnitude and sophistication of air tanker operations of 2002 was far more than the dauntless pilots and mechanics of the mid-1950s had envisioned when they began tanking Stearmans, TBMs, and Twin Beeches. The air tanker contracting process came under scrutiny by the blue-ribbon panel, which concluded: "Government contracts for air tanker and helicopter fire management services do not adequately recognize business and operational realities or aircraft limitations. As a result, contract provisions contain disincentives to flight safety." One such disincentive in contracts of that

Reminiscent of a scene repeated thousands of times at Tempelhof Aerodrome in Germany during the famed Berlin Airlift of 1948, this converted Douglas C-54 Skymaster roars in on final approach to Fox Field at Lancaster California. The World War II-era transports were converted to passenger and freight liners after the war, and also made ideal fire fighting aircraft. (Author photo)

era was a financial penalty assessed for periods when aircraft were unavailable for work, which could be interpreted as inadvertently steering operators toward flying aircraft that might be better placed into maintenance status.[155]

Training and cross-tell opportunities were found by the panel to be inconsistent. "There were no contract incentives found that would encourage operators to retain full-time safety officers to supervise and manage aircrew training." It appeared as if the complexity, sophistication, and demands placed on air tanker operators had increased over the years, with no corresponding changes in contracting procedures to accommodate these changes.

The composite picture of air tanker crews that emerged from the blue-ribbon panel credits them with courage, passion, and resourcefulness. "From presentations to the panel, it was apparent that pilots are highly experienced, confident, hard-working, exceptionally mission driven, and passionate about their aircraft and capabilities." The panel's report noted: "Air tanker pilots convey a spirit of invincibility." The risks associated with barreling down on a wildfire in a mountain canyon can become addictive. "Risk is, as some indicated to the panel, one of the most attractive features of the mission."[156]

Stemming from the blue-ribbon panel's findings in 2002, the Forest Service and the Department of the Interior (DOI) announced on 10 May 2004 that they were "terminating the contract for 33 large airtankers due to concerns over the airworthiness of the aircraft and public safety."[157] The decision proved contentious as some state officials and air tanker operators considered the move excessive and unnecessary; other government officials said it was overdue. Gone were the PB4Y-2 Super Privateers of Hawkins and Powers; gone were the postwar C-130As. Aero Union sold its C-54 air tankers to a Canadian cargo operator after 2002. Bracing for the worst, some tanker operators consumed their hard-earned cash reserves or borrowed money to conduct extensive inspections on their other aircraft in the off-season months of 2003.

By 24 March 2003, two P-3 Orions were certified by the USFS for use as air tankers, following a new inspection program developed for air tankers by the Department of Energy's Sandia National Laboratories. Sandia developed inspection plans for P-3s, P-2s, DC-4s, DC-6s, and DC-7s. The Sandia inspection

Hawkins and Powers' PB4Y-2 Super Privateer Tanker 126 features modified side windows that can hinge open to ventilate the aluminum airframe while standing alert on the hot ramp at Fox Field, Lancaster, California, circa 1999. (Author photo)

plans, paid for by the USFS and running until 2006, evaluated the various air tanker operators' maintenance and inspection procedures, and developed recommendations for enhanced procedures to enable operators to detect structural problems before they became a danger to crews and aircraft. Working with specially instrumented air tankers, the Sandia team refined estimates of how often different types of air tankers would require inspections using sophisticated non-destructive testing methods that go beyond visual inspections.[158]

It was argued that federal air tanker contracts were holdovers from decades earlier, when small companies with inexpensive war surplus air tankers could get by on a shoestring one-year contract. In fact, the 2002 blue ribbon panel said a chronic symptom of problems in the air tanker/federal interface was the tanker operators' legendary can-do attitude that was encouraged by low-dollar contracts, when the exact opposite was needed in the modern world. Contracts should reward operators who excel in maintaining their aircraft in top airworthy condition instead of merely getting by, observers noted.

The cost of tanking and operating aircraft for firefighting duty continues to rise. Around 1975, a P-2 Neptune could be bought for a little more than $8,000. After being tanked it was valued at $85,000 to $95,000. Now, P-3 Orions, valued at around $3 million, are contracted for about $1 million per season, explained Asher Williams, USFS National Aviation Operations Officer in 2006.[159]

Following changes in contracts after the 2002 blue ribbon panel report, the USFS began paying for inspections of air tankers on top of contract costs. USFS and BLM also reduced the retardant load in all large air tankers by 15 percent, and made efforts to dispatch air tankers only for initial attack, and missions in which life and property were at risk.

In late 2006, Williams said, "The aircraft right now are in better mechanical, physical condition than they have been in the 30 years I have been involved... and I get to see them all."[160]

Large air tankers were the target of an NTSB recommendation on 23 April 2004 to the Forest Service and the Department of the Interior: "Develop maintenance and inspection programs for aircraft that are used in firefighting operations that take into account and are based on: 1) the airplane's original design requirements and its intended mission and operational life; 2) the amount of operational life that has been used before entering firefighting service; 3) the magnitude of maneuver loading and the level of turbulence in the firefighting environment and the effect of these factors on the remaining operational life; 4) the impact of all previous flight hours (both public and civil) on the airplane's remaining life; and 5) a detailed engineering evaluation and analysis to predict and prevent fatigue separations."[161]

Both federal agencies responded to the NTSB recommendation by ceasing operation of their contracted air tanker fleet for the 2004 fire season. The action, thought to be draconian by some observers and necessary by others, included the provision on federal lands that "unauthorized airtankers will not be utilized." Yet if state fire agencies had responsibility on federal lands, the state could use state-contracted air tankers to fight federal land fires "so long as the state maintains operational control." Yet a federal lead plane was barred from leading "an unauthorized state airtanker."[162]

The federal action created a large initial "not approved" list of 33 air tankers, including P-3, P2V, SP-2H, DC-4, DC-6, DC-7, C-130, and PB4Y, as well as air tankers used by states. The federal rationale said: "State current contracted airtankers do not currently meet NTSB criteria..." and included some DC-6s, PBY Catalinas, the Canadian Martin Mars, "and all other makes/models (except CL215/415)." California's state fleet of S-2 Trackers and North Carolina's CL-215 were approved in the belief these aircraft met the intent of the NTSB recommendation. The Canadair CL-215 and 415 series were approved because "This type is purpose designed for the mission." Likewise, single engine air tankers—SEATs—were approved because "all SEATs are certificated and operating within the original design intent."[163] The action did not necessarily keep all these air tankers out of the sky; some flew, but not under the operational control and responsibility of the USFS or BLM.

The Forest Service hired DynCorp of Fort Worth, Texas, to inspect non-federally contracted air tankers in the summer of 2004 in an effort to return at least some of the fleet to federal service. In July, USFS and BLM announced the return of five P-3s from Aero Union into the federal fold, with more expected to return after analysis by DynCorp.[164]

By late May 2005, the BLM announced the impending return to service of several P2V Neptune air tankers following a USFS-initiated engineering study of the type. The BLM tallied the 2005 fixed-wing air tanker fleet as seven contracted P-3s, up to nine P2Vs, eight military MAFFS C-130s, six Canadair CL-215 and CL-415 air tankers, 28 single-engine air tankers, plus about 70 more on standby.[165]

The air tanker landscape may include large jet transports capable of operating within their normal design envelope while releasing huge volumes of retardant rain, day or night. And it may include, by exception, more military surplus aircraft if sufficient airframe history is available to validate the safety of these aircraft in the air tanker role. And purpose-built air tankers are an option; Canadair proved the viability of this with the utilitarian CL-215 decades ago.

VLAT—Very Large Air Tanker

The USFS, working with elements of NASA, assessed two modified wide-body jet transports, a Boeing 747 and a Douglas DC-10, to explore their suitability as Very Large Air Tankers (VLATs). The resulting summary report was issued 2 March 2009. It attempted to determine the airworthiness of both VLAT aircraft for the air tanker mission, as well as the mission compatibility of these large jetliners in the firefighting environment and flight profile. Evaluators also recommended "operational usage regimes, policies, and procedures for incorporation by [the] USFS and Department of the Interior (DOI)."[166]

The evaluation participants judged both the DC-10 and 747 air tankers to be "airworthy in the configurations under study." Long-term fatigue issues would warrant further work, but the test team "concluded that the ongoing USFS continuing airworthiness program (CAP) should enable adequate monitoring of fatigue life issues." Evaluators said: "…VLAT aircraft are probably compatible with the wildland fire suppression mission, provided that they are used to supplement other aerial retardant delivery platforms rather than replace them in all environments." The study team said it believed "extremely rugged terrain will make setting up for stabilized deliveries challenging, particularly where the pilot must judge wingtip terrain clearance while maneuvering over irregular terrain for setup. These conclusions are based on pilot comments generated during multiple simulated deliveries using high-fidelity visual simulators over various terrain types."[167]

At times the study reads like a contrast between the cautious environment of engineer evaluators and the dynamic "get-'er-done" world of classic air tanker operators. The study report used modifiers like "probably" that hedged some definitive answers: "The analysis suggests that for level or gently rolling terrain where level to slight descents (less than 6 to 7 percent) are required, VLAT-class aircraft could probably be employed with few restrictions as long as they remained above 300 feet AGL [above ground level] during the delivery. Power margins for this class of aircraft, even considering the possibility of single engine failure during delivery, may actually permit climbing deliveries over very gradual slopes of less than 3 to 4 percent grade, provided suitable egress options are available."[168]

The study said: "Usage in very steep or rugged terrain is not recommended unless deliveries can be performed with minimal maneuvering, a lead plane is available, and adequate terrain clearance is available at the wingtips as well as on centerline." The study urged, at least until a body of VLAT experience is available, that minimum terrain clearance should be 400 feet "in this kind of scenario, and a climb must be initiated before any turns." Due to the large wingspans of the VLAT air tankers tested (more than 155 and 195 feet for the DC-10 and 747, respectively) terrain avoidance, especially in turns, takes on new urgency. "Consideration should be given to making altitudes slightly more restrictive to account for the longer wingspan of these aircraft." The study noted that VLAT aircraft spend a larger percentage of their flight time at low altitudes, including sometimes-turbulent fire environments, where turbulence events can exceed the amount of turbulence experienced by wide-body airliners as they pass to higher, less turbulent cruising altitudes. This means ongoing airframe life evaluations and predictions for VLAT aircraft need to take this into account for turbulence experienced.[169]

The NASA engineers looked at the DC-10 and 747 retardant delivery systems and talked with the aircraft operators, and "found the retardant delivery systems to possess a suitable level of engineering design and redundancy."[170]

More than 20,000 gallons of retardant are tanked internally in the Evergreen 747 modification. Retardant system is pressurized, and can force fluid out at different pressure levels to meet different needs. (Evergreen International Aviation)

Boeing 747-200 (Prototype Supertanker)

At a cost to the company of more than $50 million, Evergreen International Aviation developed a four-engine 747-100 wide-body airliner (N479EV) into an air tanker capable of hauling more than 20,000 gallons. Even at this volume of retardant, the 747 Supertanker weighed less than its maximum landing weight, unlike some air tankers. This allowed the 747 to land with a full load of retardant on board if it was not dropped on a fire; no need to jettison the load before landing. The 747's retardant delivery system can force the fluid out under high pressure, or at less pressure, having the effect of rainfall. The system was intended to give the 747 a drop altitude of between 400 and 800 feet AGL, higher than older, smaller air tankers. Drop speed is around 140 knots. Typically requiring an 8,000-foot runway, the Evergreen 747 air tanker has access to 80 percent of the airports in the United States. The company says the retardant delivery system calibrates the release of fluid in a way that keeps the 747 within its g-load limitations, without a sudden upsurge on the airframe as the retardant is dropped. Another tactic touted for the 747 is night delivery of retardant, which could leverage the time when fires typically are less active and easier to suppress.[171]

The Interagency Air Tanker Board issued an interim letter of approval to use the Evergreen 747 Supertanker in March 2009. This followed the Federal Aviation Administration's awarding of a supplemental type certificate (STC) for the 747 air tanker in November 2008. An Evergreen International Aviation spokesperson characterized the huge 747 tanker: "The Supertanker will forever change the way wildland fires are fought. The plane is the first of a fleet designed to accommodate the needs of U.S. and international private and public agencies."[172]

Evergreen made plans to use the Supertanker in Australia, Western Europe, and Brazil as well as the United States. Just as pioneer air tanker operators sought the greatest utility from their aircraft, sometimes alternating between agricultural spraying and firefighting, Evergreen was at press time exploring the ability to use the Supertanker to help clean or contain environmental spills and mishaps when not fighting fires.[173]

Evergreen pioneered its pressurized retardant delivery system on a convertible 747-200 that could be reverted to transport use by removing the air tanker system. This was superseded by a dedicated 747-100 air tanker.[174]

Tanker 979, the second Boeing 747 Supertanker modified by Evergreen International Aviation, can carry enough retardant to alter the way fires are battled. (Evergreen International Aviation)

Boeing 747-200
Length: 231 ft, 10 in
Span: 195 ft, 8 in
Powerplant: 4X 54,750-lbf Pratt & Whitney JT9D
Empty Weight: 392,800 lbs
Gross Weight: 833,000 lbs
Cruising Speed: 555 mph

McDonnell Douglas DC-10-10

Innovations in air tanker capabilities include a huge wide-body DC-10 Dash-10 airliner extensively modified by a company called 10 Tanker Air Carrier based at the former George Air Force Base airfield (renamed Southern California Logistics Airport) in Victorville, California. Demonstration drops were made at the airport in March 2006, followed by work on six actual fires later that season, under contract with the California Department of Forestry and Fire Protection. The DC-10, known as Tanker 910 (N450AX), also flew on a fire in Washington State that season.[175]

Tanker 910 carries three tanks faired to the belly of the aircraft centerline with a combined capacity of 50 tons of water or retardant. Its load on fires is said to be 12,000 gallons. A computerized gravity-feed dump system, "essentially a modified and scaled-up version of the system used by Erickson Air-Crane on its S64 helicopter" allows the flight crew to adjust the opening of the tank doors to control drop rate, according to California Department of Forestry and Fire Protection information. Ground clearance for the modified DC-10 air tanker is about 15 inches. The three tanks can be filled concurrently in about eight minutes. The DC-10 changes the fire battlefield. Not as maneuverable as a

1,200-gallon S-2T Tracker also employed by California, the giant DC-10 can bring 10 times as much retardant to the fire at once.[176]

For the 2006 fire season, the State of California's Department of Forestry and Fire Protection used Tanker 910 on a call-when-needed basis, which involved an activation period of 12 to 24 hours before the big jet could be on its way to a fire. The cost for this was $26,500 per hour with a three-hour minimum per day. In 2007, the state entered a contract extending through 2009 to use the DC-10 exclusively for a 122-day season from 15 June to 15 October. This brings the daily charge down to $41,000, for an annual contract year cost of $5,002,000, plus a $5,500 charge per flight hour. Maintaining and operating a wide-body jet airliner is not cheap; neither is the cost of widespread fire damage in California's often-labyrinthine wildland-urban interface.[177]

McDonnell Douglas DC-10-10
Length: 182 ft, 3 in
Span: 155 ft, 4 in
Powerplant: 3X 40,000-lbf General Electric CF-6
Empty Weight: 240,171 lbs
Gross Weight: 430,000 lbs
Cruising Speed: 521 knots

Even through the distortion of heat waves, flap extension is readily visible in photos of the first DC-10 air tanker releasing retardant on the Zaca fire in California in 2007. More than 240,000 acres in Santa Barbara County burned in that event, which ignited on 4 July. The fire's cost was placed at more than $118 million. (Darrell Brooks)

Single-Engine Air Tanker (SEAT) Concepts

Vast improvements in the capacity and safety of turbine-powered agricultural aircraft spawned a renewed interest in the use of formerly disfavored single-engine air tankers, now championed under the moniker SEAT (single-engine air tanker). With tank capacities typically ranging between 450 and 800 gallons, SEATs were employed as a quick-reaction initial air attack tool on fires managed by the BLM.

Portable retardant plants and refueling capabilities enable SEATs to move to any airfield and broaden their base of operations over that required by heavy air tankers. As with agricultural operations, SEATs can even fly from improvised airstrips and straight roads near a fire. The BLM embraced the use of SEATs in parts of the western United States, establishing agreements with a number of airports to enable quick positioning of SEATs close to fires. The efficacy of SEATs as part of the firefighting mix may take several fire cycles to evaluate. To some observers, it is startling to visit an air tanker base and see a single-engine agricultural aircraft parked where a four-engine bomber stood alert duty previously.

The Air Tractor company, descendant of Snow agricultural aircraft, marketed the successful AT-802

With the helmeted pilot standing by in the cockpit, loaders fill an Air Tractor 802 for another sortie. (BLM)

series of SEATs. The basic AT-802 landplane model has been augmented since 2003 with the Fire Boss version, equipped with amphibious Wipaire floats and a water scooping system.

Air Tractor, Inc., of Olney, Texas, traces its roots to agricultural aircraft pioneer Leland Snow, whose S-1 and S-2 agricultural application aircraft date back to the 1950s. The Snow S-2C flew as an early single engine air tanker. Snow formed the Air Tractor company in the early 1970s. By 1990, Air Tractor developed the AT-802, an outgrowth of the earlier AT-503, for SEAT work.

The land-based Air Tractor AT-802F can tank 820 gallons. Though its heritage may be rooted in ag sprayers and dusters, 802 is "designed from the wheels up to fight wildfires," the Air Tractor company says. Able to land with its retardant tank full, the AT-802F can conserve retardant in the event of a recall after takeoff.[178]

The amphibious Fire Boss version of the AT-802 can start from a land base with a load of fire retardant. Once this initial load is discharged at the fire site, the Fire Boss can scoop water from nearby lakes or rivers and stay in the fight, loading up to 820 gallons of water in 14 seconds as it skims over the water. The Fire Boss had a "$2.5 million acquisition cost" in 2009, according to company literature.[179]

The Fire Boss uses a reversible five-blade Hartzell propeller. The aircraft can carry 74 gallons of firefighting foam additive. Unlike the landplane-only variant, the float-based Fire Boss has a maximum landing weight of 11,500 pounds; its takeoff weight is 16,000 pounds. One Canadian team used two landplane 802s in concert with one Fire Boss on a fire. The faster landplanes arrived first, dropped their retardant, and returned to base while the Fire Boss, third on the scene, dropped its initial load of retardant and then remained in the vicinity, scooping water to drop on the fire until the return of the landplanes. Typical Fire Boss operational tactics envision scooping 650 gallons the first time, due to fuel weight restrictions, gradually increasing to 800 gallons as fuel weight is relieved. Thousands of

The era of the SEAT (single engine air tanker) is epitomized in this view of an Air Tractor 802 dropping retardant. Readily deployable, SEATs add to the versatility of the air tanker fleet. (BLM)

gallons of foam-treated water can be delivered by a single Fire Boss if a navigable water source is within reach of the fire site.[180]

Air tanker operations, stereotypically a phenomenon of the North American continent, are in reality global. The Republic of Croatia Defense Ministry used five AT-802 SEATs, including some Air Boss amphibious versions. In the 2008 fire season, Croatia deployed its Air Tractors on patrols, training, and firefighting sorties along the Croatian Adriatic coastline, responding to at least 61 wildfires, and dropping an aggregate of 23,000 tons of water and fire retardant.[181]

More than 20 Fire Boss SEATs operated in other parts of Europe beginning in 2006, logging time in Spain, Portugal, Italy, and France. During that time, six were active in Canada. On 19 April 2008 the Fire Boss received federal certification—carding—as an air tanker that could be contracted to fight fires on federal land.[182]

Thrush Aircraft, Inc., of Albany, Georgia, produced the turboprop 660 Turbo Thrush agricultural aircraft in a SEAT-capable variant in 2005. The Thrush is an inheritor of several generations of ag aircraft companies including North American Rockwell (Snow Aeronautical) and Ayres Corporation. Marsh Aviation of Mesa, Arizona, previously engineered a Turbo Thrush agricultural variant in the 1970s.

A Grumman design, the ruggedly utilitarian Ag-Cat biplane was subsequently produced by Schweizer, with subsequent assembly and partial manufacture undertaken by Ethiopian Airlines for much of the market in Africa in the 1990s. First flight of the Grumman design was in 1957.[183] As with other agricultural aircraft, the Ag-Cat underwent engine changes including conversion to PT6 turbine power. A number of Ag-Cats have been used as single-engine air tankers in addition to their original primary mission of agricultural spraying.

Mid-Continent Aircraft Corporation of Hayti, Missouri, offered in the 1980s a 500-gallon tank system for firefighting with its King Cat, a re-engined Ag-Cat powered by a Wright R-1820-202A radial engine delivering 1,200 hp.[184]

In the 1970s Poland's Pezetel (PZL) company developed the M-18 Dromader, a low-wing single-engine agricultural aircraft that lent itself to the SEAT concept. The M-18 could be fitted with a Transland gate on the hopper giving a drop opening of 390 square inches, or a Melex gate with an opening of 656 square inches. The larger Melex gate promoted a flow rate of

AT-802F Fire Boss
Length: 35 ft, 7 in
Span: 58 ft
Powerplant: 1X Pratt & Whitney PT6A turboprop
Empty Weight: Approximately 8,300 to 16,000 lbs
Variable Gross Weight: 16,000 lbs
Cruising Speed: 150 mph

Grumman Ag-Cat
Length: 27 ft, 7 in
Span: 42 ft, 4.5 in
Powerplant: 1X Wright R-1820
Empty Weight: 4,816 lbs
Takeoff Weight: 6,300 lbs
Cruising Speed: 135 mph

Melex M-18 Dromader
Length: 31 ft, 1 in
Span: 58 ft, 1 in
Powerplant: 1X PZL Kalisz ASz-62IR radial engine
Empty Weight: 5,975 lbs
Takeoff Weight: Approximately 10,000 lbs
Cruising Speed: Approximately 120 mph

AT-802F
Length: 37 ft, 6 in
Span: 59.2 ft
Powerplant: 1X Pratt & Whitney PT6A turboprop
Empty Weight: 7,210 lbs (with spray equipment)
Takeoff Weight: 16,000 lbs
Cruising Speed: Approximately 150 mph

660 Turbo Thrush
Span: 54 ft
Powerplant: 1X Pratt & Whitney PT6A turboprop
Takeoff Weight: 14,150 lbs
Cruising Speed: 185 mph

Piper Pawnee-B
Length: 24 ft, 7 in
Span: 36 ft, 2 in
Powerplant: 1X Lycoming 0-540
Empty Weight: 1,488 lbs
Gross Weight: 2,900 lbs (maximum takeoff and landing weight)
Cruising Speed: 100 mph

450 gallons per second; the smaller Transland opening yielded flows averaging 80 gallons per second. The Dromader carried 500 gallons.[185]

Dromader M-18s as air tankers came to Canada in 1980. By 1984, six were in service in the province of New Brunswick. A Canadian evaluation of Dromader drop characteristics that year found the aircraft valuable to Canadian wildfire efforts. Sometimes this efficiency was enhanced by the knowledge of a pilot who could vary the amount of overlap from one release to the next to maximize retardant properties, depending on the kind and magnitude of the fire and its fuel. In some forests, the first release coated the canopy of trees, and only subsequent overlapping releases penetrated to the ground. Short travel times boosted the ability of air tankers, especially SEATs like the Dromader, to provide optimum retardant coverage while a fire burned. Aircraft like the Dromader can sometimes be moved closer to the fire if support equipment can accompany them, since these small, rugged agricultural aircraft can fly from paved or unpaved surfaces.[186]

Melex, an American subsidiary of PZL, evolved a PT6A turbine-powered version of the Dromader in the mid-1980s. An amphibious version was intended as a water scooper for firefighting. Water, untreated or foam-treated, could be dropped.[187]

Adaptations of Piper's single-engine low-wing Pawnee crop duster saw service as single-engine air tankers as far back as the 1960s. Oklahoma State University (OSU) fostered tests of Pawnees in 1963 in an effort to control range fires—grass fires—a prairie equivalent to the rampaging forest fires of the western, eastern, and southern United States. The Oklahoma retardant applications favored one part of liquid 11-37-0 fertilizer to five parts water. In a promotional brochure, OSU enthused: "Almost any plane used for crop dusting can be used if it has a controlled dump system."[188] Agricultural tank capacity of the Pawnee was about 150 gallons.

Interagency Air Tanker Board

The Interagency Airtanker Board (IAB) evaluates aircraft and tanking systems as a complete aerial delivery system, in an effort to determine suitability for the air tanker role. Reporting to the National Interagency Aviation Council, with a masthead that features the logos of six agencies (BLM, USFS, National Park Service, Bureau of Indian Affairs, U.S. Fish and Wildlife Service, and the National Association of State Foresters), the IAB is the authority for all air tankers approved for use on fires in jurisdictions covered by the six participating agencies. IAB membership includes three appointees by the USFS, three by the Department of the Interior, and a chair appointed by the National Interagency Aviation Council. The IAB is chartered to "administer an evaluation and testing process to determine acceptable aircraft, tanks, and gate system

Tanker 123 brings relief on the Jackrabbit fire in Nevada on 10 July 1994. (Keith Conrad)

performance. The Board shall review and evaluate proposals for new or modified aircraft and approve or reject such for use by the cooperating agencies." The IAB seeks to "promote and foster general improvement in retardant delivery systems and airtanker safety." The IAB charter spells out: "The Board shall promote standardization and cooperation among the using agencies and the industry in all aspects of aerial delivery systems utilization."[189]

In an effort to standardize air tanker specifications, the IAB devised parameters for fixed-wing air tankers to operate under. Amphibious aircraft used as water scoopers must be able to take off with a load of water or retardant, and carry a fire suppressant additive sufficient for at least 10 more drops at a 0.3 percent concentration. The scooper must tank enough fuel in this configuration for two and a half hours of operations at 55 percent power in cruise at 5,000 feet mean sea level (msl), and the aircraft must have a load limit factor of not less than 2.5 positive and 1.0 negative gs in cruise flight. The scooper must, after picking up a full load of water on the run, be able to climb at a rate of at least 200 feet per minute (fpm). Longitudinal stability must be positive in the drop configuration throughout the drop speed range. The IAB specification requires: "Release offload in all normal drop configurations and at all normal drop speeds does not result in dangerous or seriously objectionable flight characteristics." Flow rates of water or retardant "shall be at least 400 gallons per second." The IAB does not evaluate single-engine scoopers, so the requirement is for multi-engine aircraft in this category.[190]

For multi-engine land-based air tankers, the IAB grandfathered-in aircraft qualified before 2006; air tankers submitted for evaluation

Privateer Tanker 84 (the second to be registered N6884C) flew with Rosenbalm Aviation out of Medford, Oregon, when photographed on 13 April 1966. Bomb aiming station and faired-over bow turret location clearly show the patrol bomber pedigree of this Privateer. Later, as a Hawkins and Powers Super Privateer, N6884C received a custom canopy, B-25 engine packages, and a revised nose using an upended jet canopy for forward visibility. That's the author, at age 15, inspecting the cockpit. (Carl M. Johnsen)

Facing the late afternoon sun, four large air tankers pause during heavy fire action in Southern California circa 1999. Nearest the camera is Hawkins and Powers' Super Privateer Tanker 123, fitted with a Coast Guard observation nose. This aircraft was later lost fighting fires in 2002 in Colorado. Privateer Tanker 126, next in line abreast, features Hawkins and Powers' innovative use of upended jet canopies mounted to the noses of Privateers to give the pilots forward and downward visibility. Lockheed P-2 Neptune sits between Tanker 126 and the last aircraft in the row, a Douglas C-54 transport. The variety in this photograph speaks volumes about the ingenuity air tanker operators employed to place a wide variety of available surplus aircraft into service protecting America's wildlands. (Author photo)

beginning in 2006 needed to meet specific requirements including the same Federal Aviation Regulations (FARs) as scoopers. Performance parameters for land-based multi-engine air tankers were set higher than for scoopers: "The proponent shall demonstrate that the aircraft is capable of being operated at the Board-approved maximum operating weight in accordance with operating limitations imposed by the applicable (type certificate) or (supplemental type certificate) during all approved operations. Board-approved maximum operating weight shall include the following:

- Empty weight in airtanker configuration.
- Minimum required crew and flight kit.
- Full retardant payload, being the maximum payload for which the proponent seeks approval.
- Two and one-half hours of fuel computed at 85 percent of maximum cruise speed at 15,000 feet msl (mean sea level) without exceeding manufacturer or FAA operating limitations.
- All other necessary fluids, i.e., engine oil, hydraulic fluid, water, methanol, etc."[191]

The aircraft should have a limit load factor of not less than 2.5 positive gs and 1.0 negative gs in cruise flight. With no wind at sea level, and a computed temperature of ISA +30 degrees F, the aircraft shall have a takeoff ground roll not to exceed 6,000 feet at maximum operating weight. From this basis, the aircraft could be loaded lighter as required to ensure it could maintain a climb rate of at least 100 fpm with a retardant load on board while sustaining a critical engine failure at lift-off. Jettisoning the load, the aircraft must be able to exceed all climb gradients specified in FAR 25 with a critical engine failure at lift-off. As with scoopers, multi-engine land-based air tankers must ensure that "release of retardant in all normal drop configurations and at all normal drop speeds does not result in dangerous or seriously objectionable flight conditions." The minimum drop speed cannot be less than minimum control speed or stall speed in the drop configuration, and the maximum drop speed cannot exceed the design maneuvering speed of the aircraft. Longitudinal stability must be "positive at drop configuration throughout the drop speed range." The IAB specifies that such tankers "shall be capable of descending at Board-approved maximum operating weight along a 13 percent (7.41 degree) slope for 1 minute and leveling off at 3,000 ft pressure altitude in the drop configuration without exceeding maximum drop speed."[192]

The use of existing airframes versus new production continues to be a topic for consideration in future air tanker development. The IAB requirements address the issue: "Aircraft proposed shall be in current production, or available in sufficient quantities, to support twice the proposed number of flyable air tankers." (Air tankers approved before 2006 only need to be "available in sufficient quantities to support the proposed number of flyable air tankers.")[193]

Cockpit field of vision, a topic of discussion in early TBM Avenger evaluations, and evident in the efforts of PB4Y-2 Privateer operators to eliminate a number of the typical Privateer cockpit canopy bars going back to the 1960s, is addressed by the IAB. The field of vision is to be unobstructed, and quantified with a laser visibility evaluation device.

Busy fire action as B-17 Tanker 89 taxies, while a DC-7 carrying a long ventral tank takes off overhead at Chico, California, in 1984. (Donald Keller)

The IAB set evaluation standards for SEATs carrying 800 gallons or more, and certificated under FAR 14 CFR 21. For SEATs, the IAB-approved maximum operating weight must include the aircraft empty weight in air tanker configuration, minimum required crew and flight kit, full retardant or suppressant payload, 90 minutes of fuel at average burn rate, and all other necessary fluids. The limit load factors of 2.5 positive and 1.0 negative gs in cruise flight apply to SEATs. And, as with other tanker types, SEATs must show positive longitudinal stability in the drop configuration throughout the drop speed range, and the release offload under all normal conditions of configuration and speed "does not result in dangerous nor seriously objectionable flight characteristics." Flow rates must meet at least 400 gallons per second.[194]

More than the airplane is evaluated by the IAB. Tanks and gating systems must pass tests as well. The desired certificated volume of retardant for a tank, represented in an equal volume of water, is loaded into the tank, which must remain loaded for one week without leaking more than 3½ gallons (1/2 gallon in 24 hours). The IAB has other requirements for tanks: "All tanks are equipped with an independent controlled and operated emergency dump system enabling the entire load to be dropped in less than six seconds. This system may use only mechanical, pneumatic, or hydraulic pressure for operation. Emergency systems operated by pneumatic or hydraulic pressure are isolated from the normal tank and gating system pressure. Normal function or failure of the normal system pressure does not affect the emergency system pressure." Normal opening of tank doors is to be via "primary switches or other mechanism located on the control yoke or stick."[195]

All tanks are to have the ability to be offloaded through standard 3-inch couplers. When offloaded in this way, IAB evaluations require no more than 7 percent of the total approved capacity for retardant can remain in the tank once offloading is complete. Tank drop doors must be capable of closing in flight: no provision for one-way drop lanyards in the modern air tanker fleet. Tanks must be capable of filling at not less than 500 gpm through a 3-inch-diameter hose, connected to either side of the aircraft or at the tail. (Filler locations can determine how safe it is to "hot load" an aircraft with engines running.) Tanks must have adequate cross-flow so that the fluid will be level in all compartments within 30 seconds after the loading pump is stopped. The IAB further specifies: "All tanks with compartments which can be sequenced individually in the normal drop configuration are constructed so as to eliminate leakage from one compartment to the other when one of the compartments has been evacuated... Leakage from adjoining tanks producing a combined flow of greater than 1 gpm will constitute leakage within the means of this paragraph."[196]

Wildfire studies indicate retardant coverage levels adequate to suppress typical forest and range fires vary between 1/2 and 10 gallons of retardant per 100 square feet. Fuels, weather, and fire behavior are all variables in these computations. The majority of fires have a narrower band of adequate retardant coverage, ranging from 1 to 4 gallons per 100 square feet. Air tankers are tools to quickly build a fire line around a blaze, using retardant to hold the flames until ground fighters can secure the area. The IAB standards note, "...the line building efficiency of individual airtankers is increased by incorporating the ability to regulate the flow rate of the retardant during release at different release volumes." Maintaining a constant line of retardant is accomplished two ways. In conventional systems, the tank is divided into several compartments that can be released sequentially to produce the desired fire line effect. Non-conventional systems use computerization to vary the opening of the tank to ensure constant flow throughout the drop, as set by the crew in advance.[197]

For conventional tanks, the smallest compartment size is specified as 200 gallons. Conventional tanks must be capable of more than one release, with a formula set by IAB standards. Air tankers of 800- to 999-gallon capacity must be able to make two or more equal-volume releases; tank systems holding 1,000 to 2,499 gallons must be able to make four or more equal-volume releases; and aircraft tanking 2,500 gallons or more must be able to make six or more equal releases. This allows a tanker to make multiple drop runs before returning to the tanker base to reload, giving fire managers greater flexibility. Conventional tank multi-tank drops must demonstrate flow rate consistency throughout the drop to ensure the retardant line will not have gaps in coverage. The amount of retardant fire line an air tanker can lay down changes dramatically with the coverage required for specific fire conditions. At minimal coverage of a 1/2 gallon per 100 square feet of ground, a release of 100 gallons of retardant can lay a fire line 125 feet long. At a high coverage rate of 8 gallons per 100 square feet, at least 800 gallons are required to make a line only five feet long.[198]

Non-conventional tank systems must be able to make the same number of equal-volume releases as conventional tank systems based on total capacity. Since some non-conventional systems are pushing the volume higher, air tankers that can carry 4,000 or more gallons of retardant must be capable of making eight or more equal-volume releases. When an air tanker has multiple tank compartments fore and aft and side by side, drops of multiple compartments should select tanks that are in line fore and aft, rather than side by side or separated laterally by any other tanks. This produces the likeliest continuous line of retardant coverage on the ground.

Air tankers use vents to quickly backfill tanks with air as retardant escapes through the drop doors or gates. Inadequate venting can cause hesitation in the flow, sometimes called "Coke bottling" that can result in spotty ground coverage. The proper amount of venting is subject to variables including the amount of air space above the retardant at the start of release. Vents must be constructed to prevent retardant sloshing out or leaking during taxi and takeoff. Tanks must also have gauges to accurately measure the quantity of retardant being pumped aboard the aircraft. This is a key safety consideration to keep from overloading a tanker, especially when flight conditions dictate carrying less than a full load. Gauges also help reduce the likelihood of retardant spillover, which can lead to costly hazardous material clean-up chores. With its clean-sheet specifications for air tanker designs to be approved beginning in 2006, the IAB set out to get the most efficient tank systems possible. A safety consideration is the need to purge retardant from the fuselage of an air tanker in the event a tank ruptures, to avoid unsafe center-of-gravity shifts caused by migrating retardant inside the fuselage.[199]

Door geometry affects drop patterns. The IAB specifies: "The most desirable combination should minimize the fluid frontal area and not significantly deflect the fluid... Thus a long, narrow tank is preferred over a square tank..." Computerized drop systems help standardize and customize the volume of retardant released. The most sophisticated systems incorporate real-time environmental and flight information collected from onboard sensors or pilot input to determine optimal drop characteristics.[200]

Stylized striping reminiscent of an airliner creates the desired contrasting colors on Hawkins and Powers' Super Privateer Tanker 121 photographed at Fox Field, Lancaster, California, circa 1999. Using art, and perhaps some good humor, Tanker 121 carried a variety of color schemes as an air tanker, including a stint with large polka dots on its white base coat. Dark gray swath on lower aft fuselage is paint protection for the path of retardant. This aircraft, Navy Bureau Number 66302, is a former hurricane hunter and Coast Guard search aircraft. (Author photo)

Overheard and Overseas

Air tanker ideas that have not borne fruit include the proposal to convert A-10 Thunderbolt II ground attack jets. The rationale is that these rugged warplanes with redundant safety systems were designed for carrying loads while aggressively maneuvering near the ground. Obstacles to the plan include the likelihood that the DoD will not release any A-10s.

Argentina's CATA airline designed a removable payload tank for the bulky bullet-shaped Israel Aircraft Industries Arava twin-engine transport turboprop in 1980. Said to deliver 6,000 pounds of liquid in four seconds, the tank was loaded through the Arava's swing-open tail cone.[201]

In Russia, a jet transport (Ilyushin IL-76) contemporary to the Lockheed C-141 Starlifter has been tanked with a simple system using a pipe with a vent in front and a gate in the rear, activated by simply pulling a lever to release 10,000 gallons. The system requires a positive deck angle—nose high attitude—to operate. The IL-76 air tanker was under contract in Greece at least one year. Not certificated, it is unlikely it will be seen in the United States.[202]

A year after the two fatal inflight break-ups that altered the American air tanker scene drastically, the USFS in 2003 was reportedly considering raising a fleet of government-owned Lockheed S-3 Viking antisubmarine jets to be operated by contractors. The twinjet Vikings to be retired by the U.S. Navy still had substantial airframe life remaining, and were designed for low-level maneuvering and abrupt aircraft carrier landings, suggesting they could beat the odds as air tankers.

NASA theorized that software could be perfected in the future to inform a swarm of small UAVs—uninhabited aerial vehicles—acting as air tankers to flock to a fire and release retardant. Using software that mimics the abilities of flocks of birds and schools of fish to operate in close proximity while anticipating and avoiding collisions, this free-thinking concept might one day add a high-tech tool to the arsenal of firefighting options.

In 1971, Lockheed proposed a modular firefighting system the company called FireStar that was to have been compatible with the C-130. FireStar was envisioned to use either a 4,000-gallon tank, or alternately, "a number of small containers (up to 222) released into the air by parachute extraction, by conveyors, or by gravity with a drop rack arrangement similar to those used for naval depth charges," according to a news release. "Drums containing the retardant could be set up to burst at some preselected altitude or simply on ground impact."[203] If FireStar did not revolutionize the air tanker world, it did show the emphasis one major contractor placed on the need for periodically rethinking air attack concepts.

Also Tanked

The search for viable aircraft that could be converted to the air tanker role produced some one-off and briefly used machines that did not catch on. At least one single-engine Douglas Dauntless dive bomber, an Army A-24 variant, was flown as an air tanker by Aero Flight of Troutdale, Oregon, in 1960. Registered N4488N, this Dauntless had an external retardant tank.[204] Primarily used as a spray plane for mosquito abatement, this A-24 later saw restoration to Navy SBD standards, passing through civilian owners and going to the U.S. Marine Corps Museum at Quantico, Virginia.

Two or three North American AT-6 Texan trainers, smaller and lighter than the Oregon Dauntless, were tanked around 1961 to hold 220 gallons. Two crashes of AT-6 fire bombers did not contribute to longevity for the type as air tankers.

Versions of the Twin Beech (military C-45 transports and AT-11 bombardier trainers) were tanked in the early days, showing up in places as diverse as California, Nevada, and Florida.

A single Grumman (Columbia) J2F-6 Duck amphibian, Tanker C-07, operated out of Arizona circa 1963. It was capable of scooping 300 gallons. That year, an uprated Vultee BT-13 Valiant trainer (N57282), powered with an AT-6 engine, flew with a 300-gallon tank capacity.

The air tanker community made two efforts at converting versions of the Northrop P-61 Black Widow, a World War II night fighter as big as a medium bomber. The first was a P-61B (N30020) of Flying Farmers, Inc., in Fresno. Carrying 1,600 gallons in 1963, the Black Widow crashed on a fire scene in August 1963 east of Porterville, California, killing pilot Robert Savaria. A year later, a Northrop F-15A Reporter photoreconnaissance variant of the Black Widow (N9768Z) made its debut as Tanker E-35, carrying 1,800 gallons in a bulging belly tank modification. Longer lived than the P-61B tanker, the F-15A succumbed in 1968 to a non-fatal takeoff mishap.[205]

Drop tank remnants from an Interstate TDR-1, a light twin-engine design from World War II intended to carry a 2,000-pound bomb, lingered at Tulare, California, into the late 1980s, possibly vestiges of a proposal made there in the 1960s to use a TDR-1 (NX-7790C) with an external tank of about 200-gallon capacity.

At least one Curtiss C-46 Commando transport was tanked, but evidently not used. A second C-46, flown by the Forest Service, participated in an abortive attempt in 1962 to use radio-controlled rocket-boosted gliders to deliver 100 gallons of retardant over a fire.[206]

Fallen Fliers

At least 170 air tanker crewmembers, bird dog crewmembers, and helicopter crewmembers have died in the line of duty in the United States since 1958. Working up from the early days of the business, the aircraft involved in crashes were initially overwhelmingly World War II-era surplus: at least four PBY Catalinas; 11 TBM Avengers; two or three B-25 Mitchells; six F7F Tigercats; three or four A-26 Invaders; three PB4Y-2 Privateers (with at least three more Privateers written off in non-fatal losses); several B-17s; and the occasional rarity like a P-61B Black Widow and an N3N biplane. Newer generations of air tanker losses include about seven P2V Neptunes; four C-119s; two DC-4s/C-54s; two C-130s; nine or 10 S-2 Trackers; a P-3 Orion; and an AJ Savage. DC-7s, SEATs, bird dogs, and helicopters are included in a share of the losses as well.[207]

To the north, Canadian air tanker operators have sustained their share of casualties, including the loss of a huge Martin JRM Mars (CF-LYJ, *Marianas Mars*) in June 1961 that killed four crewmembers.

Remembered by their fellows, the families of fallen aerial firefighters are aided by a fund established by the Associated Airtanker Pilots (AAP).[208]

The glassy surface of California's Lake Elsinore reflects Hawaii Mars *at rest during a lull in firefighting operations in October 2007. This aircraft was a sister ship to the ill-fated* Hawaii Mars *that crashed in June 1961 with the loss of its crew.* (Author photo)

The rigors of air tanker work included a fire that consumed Tanker 42, the former Moby Dick, *on the runway at Wenatchee, Washington, in 1972. The crew escaped. Parts were dragged off the pavement to create an impromptu ghostly tableau.* (Author Photo)

Chapter Three

CANADIAN TWIST

Spectacular, if uncommon, view shows three Canadair CL-215 scooper amphibians making a demonstration water drop. (Canadair)

Aerial firefighting visionaries were not unique to the United States. Ellwood Wilson, forester for Canada's Laurentide Paper Company, followed a common thread in seeking aerial assets initially for timber surveying, and later for fire work. Writing in 1920, Wilson said: "In 1905, when confronted by the problem of making a map and estimate of 2,500 square miles of timber limits in Quebec, the writer looked into the possibility of using captive balloons, and found that the cost of transporting materials and equipment into the woods and the limited range of country which could be photographed from one ascension would make this impracticable."[209]

With the end of American sea patrols from Halifax, Nova Scotia, after World War I, a pair of surplus Curtiss HS2L seaplanes was abandoned there, to the delight of the St. Maurice Forest Protective Association. Beginning in the summer of 1919, the Laurentide Company of Grand Mere, Quebec, operated the Curtiss flying boats on behalf of the association, flying fire patrols as well as survey missions. Powered by the durable and reliable Packard Liberty engine, perhaps the best indicator of what American industry could produce in that era of aviation growth, the seaplanes traversed a tabulated 16,000 miles over eastern Canadian forests, producing maps and timber surveys.

Laurentide pilot Lt. Stuart Graham told a magazine writer that seaplanes were ideal in the terrain over which he flew, since lakes dotted the landscape, and at 5,000 feet altitude, he was almost never out of safe gliding distance from water.[210]

Graham's ability to scout for fires was appreciated. Since much of the region is covered in deciduous hardwoods, the highest fire danger is early spring when the woods can be dry, and before the season's lush green leaves have arrived with their fire dampening effect. On fire patrol, the Curtiss provided a view of a swath from 40 to 80 miles wide.[211]

The Curtiss HS2L aircraft could carry a Johnson fire pump and 1,500 feet of 1½-inch hose "anywhere within a radius of 150 miles within 2¼ hours. This would mean that the equipment could practically always be landed within two or at most three miles from the scene of any fire occurring in the district patrolled," forester Wilson said. While extolling the durability, reliability, and stability of the Curtiss seaplanes, Wilson observed: "The only drawback to the use of the HS2L type is the expense of the fuel, 27.6 U.S. gallons of gasoline and 1.8 gallons of U.S. oil per hour being required."[212]

The Canadian province of Ontario nurtured aerial forest patrol work in the decades between World Wars I and II. By 1926, a pair of Loening Air Yacht amphibians augmented Ontario's fleet of Curtiss HS2Ls. In 1927, four diminutive de Havilland Moth biplanes were bought by Ontario for detection patrols, releasing the Curtiss flying boats to transport firefighters and equipment to fire sites. Other aircraft, including metal Hamilton seaplanes, Fairchild 71s, Stinson SR-9s, and Canadian-built Buhl CA-6s, lasted into World War II, when a few durable Noorduyn Norseman single-engine airplanes joined the Ontario effort. Convinced that aircraft were indispensable to Ontario's forest protection activities, after the end of World War II the Ontario Air Service worked with de Havilland in Canada to design an aircraft for its bush operations. The result was the reliable, and ultimately ubiquitous, de Havilland Beaver single-engine aircraft, at home on floats or wheels.[213]

In the late 1940s, it is reported Australian foresters tested 50- and 65-gallon liquid tanks dropped from a B-24H Liberator, an Avro Lincoln, and P-51D Mustangs. These tests also evaluated of the efficacy of retardants added to water, versus the use of plain water. A 1947 professional journal reported Australian P-51 Mustangs dropped 750-pound bombs on patches of blazing scrub, with results described as most encouraging.[214] Around this time forest workers in Canada rightly concluded hard-cased firefighting bombs, fused or not, weren't the answer.[215] The abundance of lakes suitable for aircraft operations in Canada pointed the way to air tankers that would take on water from a source near the fire, shortening travel time and increasing the number of drops an aircraft could deliver in a given period of time.

Early Canadian efforts centered on waterproofed paper bags, tested in 1949 by the Ontario Department of Lands and Forests. Operational debut came on 9 September 1950 when the crew of a de Havilland Beaver released nearly a ton and a half of water on a blaze north of Sault Ste. Marie. The aerial action was credited with slowing the advance of the fire sufficiently to permit ground crews to quell it. Rustic by today's standards, the Beaver's inaugural air attack necessitated landing on water to a full stop, with a crewmember climbing out on a float to scoop water into 5-gallon paper bags that subsequently were dropped on the fire.

Refinement of the technique over the next few years led to 3.5-gallon plasticized paper bags, with as many as 24 carried by a Beaver for release in groups of eight on a pass. One pass could cover a swath 10 feet wide and 90 feet long. Typically, the Beaver would land on the nearest body of water, offloading two firefighters to head overland to the fire while the pilot and an assistant filled and dropped the water bags.[216] Filled with water, the bags weighed 35 pounds apiece. The aircraft used a conveyor with rollers to launch the water bombs through a ventral camera hatch. A crewman loaded the conveyor and pushed the bombs during the attack while the pilot flew the aircraft. Bombing runs often were made at about 200 feet at just faster than 80 mph. The bombs splashed 10 to 12 feet apart, with the crew making runs from different directions to saturate the target area.[217]

The technique held promise for lightning-strike fires if detected early enough before they could bloom into major blazes.

Tanking on the Fly

Pioneer efforts at making seaplane scooper air tanker aircraft were accomplished by the Ontario Department of Lands and Forests in 1957. Starting with water tanks fitted to the fuselage of a single-engine de Havilland Otter, the Canadians were impressed with the flow of water they could produce over a fire, yet this installation required the aircraft to come to a stop for filling the tanks.[218]

The next iteration put a 90-gallon (80 imperial gallons) tank atop each float of an Otter, with a 2½-inch-diameter pipe extending into the water as a scoop. At a fast taxi, the Otter could fill the tanks in 18 seconds. The system promised to put 1,800 gallons of water on a fire in an hour, presuming the fire was within 2 miles of one of Canada's many navigable lakes. To release the water, the tanks were rotated to empty them. Perforated

Philippine Mars ***shows its side gates for water release in this view that probably doused the photographer.*** (Forest Industries Flying Tankers)

Registered CF-FEU-X, an early Canadair CL-215 salvoes water in side-by-side plumes. (Canadair)

baffles in the cylindrical tanks minimized sloshing of water.[219]

The Otter also carried fire crews and equipment to lake landings near the site of a blaze, saving hours and sometimes days over ground transportation to the same remote location. Sometimes, radio transmitters were airdropped near a remote fire to help guide ground crews the last miles to the site.[220]

The early Otter tanks represented a major step forward from previous Canadian efforts using paper or plasticized bags as water bombs. But the zenith of scoopers awaited Canadian conversions of larger PBY Catalinas, the huge Martin JRM Mars flying boats, and purpose-built Canadair CL-215 amphibious scoopers. By 1969, a Canadian engineer reported improved PBYs capable of scooping a full load in a run of only 1,600 feet, enabling a delivery rate as high as 20 loads in an hour when the body of water was close enough to the fire to permit this. Turbo Beavers and Twin Otters, skimming their loads in runs of only 600 feet, proved capable of delivering more than 35 loads an hour.[221]

Beaching gear bobs in the waters of Sproat Lake in summer 1967 as the **Philippine Mars** *taxies in from a sortie. A crewmember is already in the nose hatch to facilitate operations on the lake.* (Author Photo)

When retracted the scoops on **Hawaii Mars** *are protected by shrouds. Line of flight is to the right in this 1974 view.* (Author Photo)

Canadian Catalina Scoopers

By 1960, innovative Canadian firefighters dropped water from external underwing tanks hung on PBY Catalinas—Cansos in Canadian parlance. The system worked, but necessitated full-stop landings to replenish the water tanks after each sortie. Engineer J. K. Hawkshaw of Field Aviation, Ltd., had a better idea. He fitted Cansos with two internal tanks side-by-side in the fuselage near the center of gravity. An extendable probe captured water as the PBY-5A skimmed the surface of a body of water in a touch-and-go maneuver. Hawkshaw's invention served two important purposes: it lessened the time between drops, and decreased pilot workload. In Quebec by 1964, Canso scooper crews

Some PBY-5A Catalinas received aftermarket enlarged rudders. This is Flying Fireman's Tanker 3 (CF-FFZ) at Sidney, British Columbia, in 1967. (Author Photo)

Flying Fireman Tanker 5, a PBY-5A (or Canso in Canadian parlance) uses paddle props instead of slimmer toothpick variety common to PBYs. Heavy overcast above Vancouver Island suggests no fires on this day in 1972. (Author Photo)

Tanker 5 (CF-IHN), a Catalina (or Canso) of Flying Fireman, shows evidence of civilian modifications including removal of side blisters and addition of extra fuselage windows. (Author Photo)

Toothpick props, original bow, and side blisters make Catalina Tanker 7 of Flying Fireman nearly stock in this 1972 photo at Sidney, British Columbia. (Author Photo)

had a mandatory upper limit of 70 air drops a day, higher by 10 than the number settled on for the externally tanked PBYs.[222]

Total capacity of the Cansos fitted with internal tanks was 800 Imperial gallons, which could be evacuated over a fire in less than a second. Scooping a full load could be accomplished in as little as 14 seconds. Vents to the outside facilitated quick loading and prevented damage from over-filling. When extended, the filler probe was at risk from striking objects on the water's surface, and was designed to break away in a contingency. Piping from the probe to the tanks used divergent hydrodynamic engineering tenets to prevent choking in the piping. This divergence slowed the water's velocity to about a fourth of the Catalina's forward speed by the time the tanks were reached. The bomber's special dump doors each were 27 x 60 inches, and formed part of the bottom of the hull when closed. They could free-fall open, and had hydraulic jacks for closing. Typical altitude for drops was 100 feet, sometimes lower.[223]

The Province of Quebec operated seven PBYs by summer 1964, with plans to acquire more. That year, Newfoundland had four field-modified Cansos in service, Manitoba used a pair under charter agreement, and Saskatchewan and Ontario each had one under charter. About three more Canso air tankers served in France at that time as well.[224]

Mighty Martin Mars Scoopers

While Catalina air tankers were under development in eastern Canada in July 1959, a consortium of lumber companies out west in British Columbia created Forest Industries Flying Tankers (FIFT) Ltd. The group ambitiously bought the U.S. Navy's four remaining Martin JRM Mars four-engine flying boats plus spare engines, support equipment, and even a forward fuselage section from an unbuilt Mars transport. The Mars, with its immense 200-foot wingspan, roars to life with the power of four R-3350 radial engines. At a time when the capacities of other nascent air tankers were measured in two- or three-digit numbers, the Mars could deliver as much as 7,200 U.S. gallons.

Martin JRM Mars Air Tanker Identification

USN Bureau Number	Nickname	Canadian Civil Registration
76820	*Philippine Mars*	CF-LYK
76821	*Marianas Mars*	CF-LYJ
76823	*Hawaii Mars*	CF-LYL
76824	*Caroline Mars*	CF-LYM

The *Marianas Mars* was the first JRM to be converted. Near Victoria on Vancouver Island, Fairey Aviation of Canada, Ltd., installed four plywood and Fiberglas tanks in the hull, with a pair of scoops for on-the-run loading of 7,000 gallons of water. The effect was remarkable: a swath 175 feet wide and 600 feet long could be covered in a single pass. First tested 6 March 1960, *Marianas Mars* went to work on fires that September. The young FIFT effort suffered a severe setback 23 June 1961, when *Marianas Mars* crashed while fighting a fire on Vancouver Island, killing all four crewmembers. A second blow came on 12 October 1962 when the beached *Caroline Mars* was savaged by hurricane-force winds that hit that portion of Canada and the northwestern U.S. Parts were salvaged from *Caroline Mars*, but the hulk was scrapped by 1967.

Open hatch on side of fuselage near the nose of Hawaii Mars provides crew access from a small boat that would deploy from the shoreline whenever a fire call arrived. (Author photo)

On a hot August day in 1967, Hawaii Mars *made an inland demonstration drop over the Abbotsford International Air Show in British Columbia.* (Author Photo)

With a remote flight engineer's station tracking Mars' vital signs, the flight deck of Hawaii Mars *looks austere and tidy.* (Author Photo)

Hawaii Mars' *flight engineer's station October 2007.* (Author Photo)

Clean and functional, the interior of Hawaii Mars *includes catwalks to the extreme aft end of the fuselage, as photographed in October 2007.* (Author Photo)

Reinforced bow structure is evident in internal view of Hawaii Mars *while at Lake Elsinore, California, in 2007. Removable hatch facilitates mooring operations.* (Author Photo)

FIFT continued conversion of the remaining two JRMs, *Philippine Mars* and then *Hawaii Mars*. *Philippine Mars* releases its water load through modified cargo doors in the side of the fuselage much as *Marianas Mars* had done; *Hawaii Mars* has drop doors in the bottom of the hull, said to deliver a more concentrated pattern when water is dropped. *Philippine Mars* has a listed total for water capacity of 4,500 Imperial gallons; *Hawaii Mars* can tank 6,000 Imperial (7,200 U.S.) gallons. The two surviving Mars flying boats have been a mainstay of fire suppression activity in British Columbia ever since, and have ranged as far as Southern California on contract.

The Mars can drop fresh- or saltwater, and freshwater can be treated with a gelling agent that allows drops from higher altitudes. The JRMs' home base on Vancouver Island's Sproat Lake is a clean body of freshwater free of the aggravations of milfoil, an easily propagated freshwater plant. When returning from making pick-ups on other lakes, Mars crews will make a scooping run in the brine of the Pacific Ocean, dropping it before returning to Sproat Lake in an effort to keep milfoil out of home waters.

When planning fire sorties, Mars crews figure fuel consumption at 450 gallons an hour in cruise and 850 gallons an hour during operations. The five member companies of FIFT were reminded to take advantage of the Mars' capabilities early in a fire's life cycle. "The Mars tanker, in common with all aerial water bombers, is most effective on relatively small fires," a 1966 FIFT Fire Control Manual advised.[225]

With thousands of gallons of water, possibly laced with additives, hurtling toward the earth, Mars water releases come with their own kind of peril to firefighters on the ground. A 1966 FIFT manual explained: "Ground crews are to be warned of tankers' approach, and if possible move quickly out of target area. If you cannot move out of target area, toss fire tools away on downhill side, lie down, bury face in arms, hard hat on head, and face the direction from which the plane is coming. If in timber, stay clear of dead snags, tops, and limbs in drop area. Material such as rocks, live and dead growth, and rolling material on slopes are particularly hazardous."[226]

The majestic sight of a red-and-white Martin Mars seaplane releasing tons of treated water on a wildfire is unmatched. The longevity of these two flying boats from another era attests to their care and handling over more than four decades of fire operations.

Twin Otter

When de Havilland went into production of Twin Otter turboprop transports in 1966, the ability of this larger twin-engine aircraft to operate as a floatplane opened up possibilities for another air tanker scooper, larger than the Beaver and single-engine Otter, yet smaller than the PBY and JRM Mars. Canada's Field

Stout beaching gear allows seaplanes like the Martin Mars to be towed on land. The beaching gear units float, and are taken out to the Mars on the water. At Port Alberni, British Columbia, in 1974, air tanker **Hawaii Mars** ***receives preparation for repainting while sister ship*** **Philippine Mars** ***floats nearby. Both began their air tanker careers in the early 1960s.*** (Author Photo)

Aviation, Ltd., built tanks for the Twin Otter's floats, which were made by Aircraft Products Limited of Vancouver, BritishColumbia. The Twin Otter could tank 450 big Imperial gallons that could be thickened with Gelgard. Doors on the float tanks were boosted on opening, speeding the exit of the water, said to improve the Twin Otter's drop pattern. A landplane variant of the Twin Otter also pioneered a 400-Imperial-gallon membrane tank delivery system, in which the tank's disposable vinyl bottom was cut by mechanical knives. The speed of the cut could be controlled to affect drop pattern.[227]

Purpose-Built Canadian Scooper

The characteristically Canadian tradition of seaplane scooper firefighting aircraft fostered a discussion in winter 1963 about the ideal attributes of such an aircraft. An early paper design, the twin-boom CL-204 floatplane, was discarded as too restrictive. The result was a clean-sheet design by Canadair, the CL-215, spanning nearly 94 feet and propelled by a pair of Pratt & Whitney R-2800 radial engines. A load of 1,200 Imperial gallons of water could be scooped in 16 to 20 seconds by early versions; later models cut this time to

Hawaii Mars ***releases water over a Canadian forest. Its drop doors are on the bottom of its mammoth hull, unlike the side exits for*** **Philippine Mars.** (Forest Industries Flying Tankers)

Hawaii Mars ***skims Lake Elsinore during fire operations in October 2007.*** (Author Photo)

Fire operations off Lake Elsinore in 2007 included sharing airspace with local birds. (Author Photo)

Canada's thousands of lakes promote the use of floatplane air tankers like this Twin Otter displayed at Abbotsford, British Columbia, in August 1969. (Author Photo)

10 seconds with an improved scooping system. From the outset, the CL-215 was designed pragmatically to fight fires and perform other utility tasks at cruising speeds around 180 mph or less.[228]

The resulting aircraft has a boxy, yet hydrodynamically efficient fuselage mated to a non-tapering high wing with ample flaps and no dihedral, followed by a large cruciform tail. The tail surfaces made use of honeycomb panels, known for their lightweight strength. Exposed mainwheels retract in the fashion of Catalinas or Grumman amphibians. A fully enclosed nosewheel extends to keep the CL-215 in a level tricycle stance on the runway. Early development of the CL-215 used engine and nacelle packages from the Douglas A-26 Invader attack bomber.

Two scoops mounted behind the step in the hull feed two internal tanks, the upper portions of which were designed to be removable to permit use of the aircraft for other purposes. The CL-215 can function as a traditional land-based air tanker, loading fire retardant from hoses, or it can tank water on the run as a seaplane. Fire retardant additives can be mixed during scooping. The efficacy of a fire suppression load on the ground can be lost to heat dissipation in 15 minutes, making turnaround times shorter than that crucial to best fire management techniques. The proximity of navigable water can make this possible. The crew of a Manitoba-based CL-215 made news in the summer of 1978 when they logged 160 attacks on fire in one day, dropping almost 1,000 tons of water total.[229]

Following extensive wind tunnel and water tank model testing, the first CL-215 was built, flying (from a runway) for the first time on 23 October 1967. It would not fly from water until 1 May 1968. The first three aircraft off the line each contributed to the

CL-215's test and certification process. Along the way, the rudder was extended, along with other modifications. Six, and later eight, flexible fuel cells in the wing give the CL-215 a range of more than 1,400 miles—effective for ferrying, but much more than a typical fire sortie requires. The ability to tank fuel permits the CL-215 to make many short-range drops before requiring additional gasoline.

Firefighters in Canada, France, Spain, Greece, and Venezuela have operated CL-215s. A pair of Thai CL-215s has seen service in patrol and search-and-rescue work, as have a number of Spanish versions. Canadair tallied 125 CL-215s built with R-2800 engines before making a change to turboprops, initially designated CL-215T.

The CL-215T could be derived in two ways. The manufacturer offered new-build propjet aircraft, and conversion kits were sold to permit conversion of piston-powered CL-215s. Canadair forecast that approximately 44 percent of the turboprop CL-215s would be used as air tankers or spray planes, with more than half expected to be employed in other roles including search-and-rescue missions as well as utility transport. The first CL-215T, a company conversion from a piston-engine model, made its initial flight 8 June 1989.[230]

The successful CL-215 airframe received an upgraded and air-conditioned flight deck with the turboprop model, increased allowable weight after taking on water, a larger four-tank water drop system, wingtip endplates, and provisions for underwing hard points and power-boosted ailerons.

Mid-span auxiliary vertical fins on the horizontal stabilizer were added after the prototype flew. Along with a streamlined bullet shape at the cruciform tail juncture, these devices improved longitudinal and directional stability that had been affected by the new thrust line of the turboprop engines.

By 1991, Canadair differentiated new construction from converted CL-215 aircraft by renaming the new turboprops CL-415. CL-415s feature improved drop patterns for firefighting. A full load can be scooped by a CL-415 in as little as 1,250 feet of skimming over water as shallow as 4 feet, 7 inches. Engines for the propjet Canadairs are variants of the Pratt & Whitney Canada PW123AF turboprop, turning four-blade Hamilton Standard propellers. The first CL-415 flew on 6 December 1993. The first customer to receive a CL-415 was France in December 1994.[231]

Canadair CL-215 CF-WYQ takes off in August 1969, showing its simple angular lines and exposed main gear retraction pockets in the fuselage side. (Author Photo)

Canadair CL-215 tanker in French registration F-ZBAX provides aesthetic view of French countryside, as well as a glimpse of its large engine nacelles perched over the wing. (Canadair)

Purpose-built Canadair CL-215 air tanker makes a demonstration drop at Abbotsford, British Columbia, on 9 August 1969. (Author Photo)

The main gear is still retracting as Skyway TBM Tanker 1 rises from the former RCAF Liberator base at Abbotsford, British Columbia, in August 1968. (Author Photo)

On Water and On Land

Canada is spangled with lakes that have shaped commerce, recreation, and Canadian life for centuries. Until 1973, the British Columbia Forest Service (BCFS) contracted amphibious Cansos (Catalinas) from Flying Fireman, Ltd., in Victoria, British Columbia, as well as land-based air tankers from Conair out of Abbotsford, British Columbia. But BCFS favored the use of long-term retardants more typically applied by landplanes than by water scoopers, and the province moved away from Cansos. Flying Fireman's Cansos could make the initial attack, taking off from their land base, carrying long-term retardant. But subsequent convenient scooper runs used water and short-term retardant that could be mixed on board. BCFS re-evaluated the use of scoopers around 1977 as costs for long-term retardant increased during the decade, making some use of short-term retardant economically viable.[232]

Flying Fireman kept its Cansos busy with contracts in other provinces, including Alberta, Manitoba, and Ontario after BCFS stopped using scoopers. Meanwhile, Conair's fleet of DC-6s, A-26s, and even single-engine TBM Avengers continued to serve air tanker requirements in British Columbia and other provinces. By 1977, Conair flew seven DC-6Bs (two strictly as large-area pest sprayers),

TBM Avenger Tankers 2, 1, and 4 of Skyway were part of a large fleet of TBMs in Canadian service when photographed at Abbotsford, British Columbia, in August 1968. (Author Photo)

a dozen A-26 Invaders, and eight TBM-3s. A Conair DC-6 provided fire protection in the Northwest Territories that year, and the TBMs could be moved where needed and configured for spraying or firefighting depending on the severity of the season. Large-area spraying for pests, including spruce budworm, was a wise hedge for Conair, and helped the company's cash flow whenever wet summers made firefighting work scarce. Also by 1977, Conair was at work on its own conversion of Grumman S-2 Trackers, envisioned as convertible tanker/sprayer aircraft.[233]

The spirit of innovation universal and vital to air tanker operators manifested itself in the unique one-off conversion of a former Royal Canadian Air Force Lancaster heavy bomber, number KB976, into an air tanker by Northwestern Air Lease, Ltd. But the roar of four liquid-cooled Merlin engines was not to be heard over Canadian wildfires as the Lancaster was sold to the historic Strathallan Aircraft Collection in Scotland after tanker trials, but before operations were begun with the iconic British bomber design. As the smallest of the western Canadian air tanker operators, Northwestern was also, by 1977, the last Canadian operator of B-25 Mitchells, flying a pair of the medium bombers long after their discontinuance in the lower 48 states.[234]

Avalon Aviation, Ltd., of Red Deer, in central Alberta, employed several Cansos for the 1977 fire season in Canada, serving contracts in the Northwest Territories and Ontario. Late in the 1976 fire season, Avalon flew a Canso to Norway to fight severe Scandinavian fires. The 1977 season included fire protection from four Cansos of North Canada Air, Ltd., based at Saskatoon for operating in La Ronge and Dryden,

Fluid hurtles earthward as Skyway TBM Tanker 1 makes a gear-down demonstration drop at Abbotsford, British Columbia, in August 1968. (Author Photo)

Air Spray A-26 Tanker 14 carried nose art depicting winged horses pulling a vintage hand-pump fire wagon. (Air Spray Ltd.)

Ontario. Government-owned Trackers were also tabbed for service in Saskatchewan that year.[235]

Air Spray Ltd., known for its bright yellow fleet of A-26s, operated at least 10 Invaders out of Edmonton, Alberta, by the 1977 season. That year, Air Spray shared air tanker duties over Alberta with Flying Fireman's Cansos, and was prepared to send as many as three of its A-26s to the Yukon for firefighting.[236]

By the last half of the 1970s, Canadian air tanker evolution was a mix of land-based and water-borne aircraft. Venerable Cansos remained in high demand in Ontario as well as the western Canadian provinces.

Firecats and Firefighters

Just across the border from Washington State, in Abbotsford, British Columbia, Conair Aviation Ltd. modified S-2 Tracker twin-engine patrol aircraft to become Firecat air tankers, and, with PT6 turboprop engines, Turbo Firecat tankers. The company also designed and fielded an air tanker modified from a turboprop F-27 airliner in 1986. The Firecat carries 870 U.S. gallons (725 Imperial gallons) of fire retardant in a fuselage tank that necessitates raising the floor by 8 inches. Larger wheels with low-pressure tires are fitted to permit soft field operations—a far cry from the S-2's original aircraft carrier deck environment. The F-27 Firefighter can carry 1,681 U.S. gallons.[237]

Conair teamed with Kelowna Flightcraft to develop an air tanker based on the Convair 580 turboprop twin-engine airliner. Using a constant-flow retardant delivery system, the CV-580 air tanker is a modern component of Canadian aerial firefighting. The 580 conversion tanks 2,100 U.S. gallons (1,750 Imperial gallons).[238]

Conair DC-6 makes spectacular dyed-water drop at Abbotsford International Air Show in British Columbia. (Author Photo)

Eleven Douglas A-26 air tankers of Canada's Air Spray Ltd. company wintered over in this photo, dormant until the following summer brought fires to the Canadian forests. (Air Spray Ltd.)

Tanker 42, a Conair Douglas DC-6, releases fluid in trail in this 1977 demonstration. (Author Photo)

All about piston power and whirling prop blades, Conair DC-6 tanker shows its tank configuration in this 1977 photograph. (Author Photo)

Chapter Four

THE MECHANICS OF AERIAL FIREFIGHTING

PB4Y-2 Tanker 85 (N3191G) in an early post-conversion configuration makes a demonstration drop in Oregon. Cockpit window configuration and nose fairing would subsequently change. (Kenn Knackstedt Photo via Donald Keller)

The romance of air attack operations is rooted in the heroic image of first-generation air tanker pilots using courage and Kentucky windage to place retardant where it would do the most good. To be sure, skill has no substitute in firefighting. But the tools have evolved far beyond a simple dump of retardant at the edge of flames. Computer-aided drop door mechanisms automatically adjust door apertures to ensure an even flow of retardant from the first to the last on a single pass. Infrared equipment, sometimes aboard remotely piloted aircraft, reveals hotspots invisible to the naked eye. And fire bosses have guide sheets to help them call for the correct amount of retardant to be released, depending on the terrain and the natural fuel that is feeding the fire.

Firefighters learned early that the volume of water or chemicals that could be carried by most air tankers was more effective at corralling a blaze and slowing its spread than extinguishing it outright. The huge Martin Mars seaplanes could release so much water with such force that it could actually pick up burning debris on the ground and push it ahead of the water, inadvertently spreading the flames; careful piloting avoided this.

Through trial and error, and deliberate experimentation, the air tanker community has amassed a body of corporate knowledge that informs modern air attack rationale.

The spectacular plume of colored fire retardant raining down on a wildfire is the obvious payoff of a choreographed operation intended to keep bombers on target. The *Interagency Airtanker Base Operations Guide* codifies this operation for a variety of users in the United States. Individual air tanker bases develop supplements to the guide to localize it where necessary.[239]

Operational Positions

Base workers are cognizant of the need to contain and clean up spills of retardant and fuel. Retardant loading pits require containment and treatment systems to treat not only spilled retardant, but aircraft wash water that may contain "metals from the aircraft, fuel, hydraulic fluid, and oils," according to the base operations guide.[240]

If there seems to be irony in the requirement to clean up retardant spills at airports, when spilling retardant at fire sites is the whole reason for being, the interagency guide does address dropping retardant in sensitive areas: "Follow guidelines indicating that retardant or foam should not be applied within 300 feet of a waterway or riparian area..." Compliance with the Threatened and Endangered Species Act is called for.[241]

The single-engine Avenger's paint may be peeling, but its engine beats out a reliable tattoo as it taxies to join a fire effort in northern Idaho in the company of A-26 Tanker A17 on 27 July 1973. Smoke in the distance is a deliberate grass fire, not the blaze to be fought. (Author Photo)

Fires quickly get named; it helps when more than one blaze occupies an area. This is the Jackrabbit fire in Nevada, tended by Hawkins and Powers' Super Privateer Tanker 123 on 10 July 1994. (Keith Conrad)

In the late-afternoon sun B-17 Tanker 16 and a smaller PV-2 Harpoon stand watch, with the remnants of burned-out Privateer Tanker 42 behind them at Wenatchee, Washington, in September 1972. (Kenneth G. Johnsen)

Hawkins and Powers Aviation Super Privateer Tanker B21 releases retardant at the edge of the smoke in this low-level drop. (USFS)

Lounging with backpacks and suitcases, a handful of firefighters at the USFS base in Wenatchee, Washington, rests in the company of air tanker warriors including PB4Y-2 **Moby Dick**, **_PV-2 Tanker 40, and the first converted Neptune in July 1970 during a lull in the action._** (Author Photo)

A-26 Tanker 57 made a demonstration drop for the camera of Kenn Knackstedt, who chronicled air tanker activities in the Medford, Oregon, area. (Kenn Knackstedt)

Air tanker bases establish a retardant jettison area that can save an outbound tanker if problems make continued flight with a retardant load unsafe, or to enable a tanker to land safely.

When loading an air tanker, mass-flow metering equipment provides the weight of retardant in pounds as it is delivered to the aircraft, regardless of the consistency of the mixture. The actual weight is crucial for each aircraft and appropriate maximums can vary with weather conditions or specific airfield operating limitations. If the mass-flow metering capability breaks down, "...the airtanker can still be loaded using the visual indicators on the retardant tank coupled with close attention to the quality control procedures outlined for the product. This procedure should only be used temporarily until the meter can be repaired or replaced."[242]

The base operations guide warns loaders: "Always notify the pilot if the aircraft is overloaded with retardant! It is critical to flight safety that airtankers are not overloaded with out-of-specification retardant that exceeds the per gallon weight limitations. Overloaded aircraft crash, it is that simple. Always determine the total weight in pounds (not gallons) loaded into an airtanker. Always notify the pilot if you have verified, or even think, the aircraft may be overloaded!"[243]

In some instances, loading is authorized while the aircraft's engines are running if all personnel involved have received hot-loading procedures training; any ground operations around an aircraft with engines turning is fraught with greater potential danger. Simultaneous loading and fueling are never allowed because of the danger for static electricity build-up.[244]

Text continued on p. 110

TANKER BASE OPS

Seeing fire bombers artfully making their drops is only part of the story. These difficult missions begin and end at the Tanker Base.

Air Tanker Base Manager

Overall boss of the base; monitors everything from fire situation to logistics and human resources; coordinates all tactical aircraft missions with local dispatch.

Retardant overflow reflects the action as a C-54 follows a PB4Y-2 toward the runway at Chico, California, circa 1984. (Donald Keller)

Ramp Manager

Oversees loading and fueling of air tankers, parking of aircraft, safety of ramp operations.

Permanent ladders leading to the overhead cockpit hatch facilitated Privateer air tanker operations. Retardant tanks in the bomb bay inhibited other quick access to the cockpit. Tanker C30 is seen at Chico, California, in 1984. (Donald Keller)

Parking Tender

This specialist verifies individual air tanker loading restrictions in consultation with the aircraft captain; clears aircraft for taxi; ensures safety of personnel working around aircraft; is liaison between pilots and providers of retardant, fuel, meals, crew rest, and other needs.

Tail-end view of a B-17G air tanker (nearest camera). The original USFS P2V-5 Neptune shows eddy effect with retardant plastering itself to the blunt aft ends of both fuselages. B-17 has retardant fill port located in end cap on former tail gun emplacement. (USFS)

Impromptu dye art splatters the loading area at Chico, California, as loaders connect two filler hoses through the tail gun emplacement on B-17G Tanker 99 during the 1984 Butte Creek Canyon fire. (Donald Keller)

Retardant Loading Crew

May be same crew as retardant mixing crew; loads retardant into air tankers and reports pounds of retardant loaded, via mass-flow meter on hose, to mixmaster after each load; verifies amount loaded into each air tanker does not exceed placarded maximum load weight listed on fuselage of air tankers, to include lesser weights as designated by the pilot; keeps ramp clean from spilled retardant.

Like spent shell casings on a battlefield, ripped and empty Phos-Chek retardant sacks litter the site of hectic mixing during a wildfire fight. Fresh bags are hoisted above the hopper to keep the battle going. (USFS)

Retardant Mixmaster

Supervises the retardant mixing crew in action; logs quantity of retardant loaded and reports this to aircraft timekeeper.

The mixmaster and the mixing crew can trace their jobs—and possibly the viability of the entire air tanker business—to a California inventor named Wim Lely, who devised retardant mixing units commensurate with the need of burgeoning air tanker operations.[248]

Retardant Mixing Crew

May be same crew as retardant loading crew; takes samples of retardant for quality control.

Clear skies behind A-26 Tanker A17 show no signs of the fire as the bomber lands at Coeur d'Alene, Idaho, in July 1973. (Author Photo)

Aircraft Base Radio Operator

Maintains communications with aircraft assigned to the base until takeoff and after landing; notifies airtanker base manager immediately of any overdue or missing aircraft; maintains log of all aircraft takeoffs and landings with estimated times of arrival and estimated times of departure.

Smoke pops away from B-25 engine exhausts on modified Super Privateer Tanker 127 at Lancaster, California, during heavy sustained fire operations in August 1997. (Author Photo)

Aircraft Timekeeper

Records all timekeeping information for each air tanker; ensures retardant use is properly documented, and landing fees are properly documented; maintains tactical aircraft use and cost information, and keeps incidents air operations staff informed daily.

Ventral filler coupling allows servicing of Douglas A-26 N4805E, Tanker 58. (Author Photo)

Normal dispatch time for an air tanker can be varied due to instrument flight conditions, air traffic, and other causes beyond the pilot's control. The base operations guide notes: "Fifteen minutes is the standard reaction time as specified in the federal airtanker contract."[245]

Federal air tanker contracts call for an equitable rotation of duties to ensure all contracted operators of large air tankers are treated equally. "In cases where state resources are operated in conjunction with federal contract items (large airtankers) on an incident primarily on federal lands, airtankers shall be rotated per the national policy with the state resources being added to the rotation after the federal resources at the beginning of each day."

Normal operations have air tankers arriving over a fire no earlier than 30 minutes after published sunrise, and not later than 30 minutes before official sunset. These times can be stretched to 30 minutes before sunrise and 30 minutes after sunset if an air tactical group supervisor, air tanker coordinator, or lead plane pilot on scene has determined, with concurrence of the tanker pilot, that visibility and safety are suitable for a retardant drop, and the dispatcher has been so notified.[246]

Hazards present themselves in a variety of ways to air tanker crews. Tanker bases post a map of known aerial hazards in the region, typically including hazards to low flight such as power lines and towers, military training routes, restricted areas and other special-use airspace, identifiable areas of turbulence, threatened and endangered species habitats, and areas of hang-gliding, soaring, and skydiving activity. To keep aircraft and aircrews prepared during periods of inactivity during the fire season, some bases have tanker crews perform water drops (instead of actual retardant) and other proficiency checks such as flying instrument approaches. Air tanker base staffs typically conduct daily pre-mission safety and operations briefings and post-mission debriefings.[247]

The ability to fight fire close to its origin includes the use of portable air tanker bases when needed to bring the basic air tanker retardant capabilities to airfields other than established permanent bases.

TBM, Inc. operated C-130A Tanker 63 (N473TM) decked out in bright yellow-and-white markings for visibility. (Author Photo)

The first Privateer to carry registration N6884C (BuNo. 66284) broke into large sections in a crash near Prescott, Arizona, summer 1959. The registration number was reassigned to a second Privateer (59701) that outlasted its multi-decade stint as an air tanker. (Ken Shake)

Retardant trails from the right tanks of a vintage Privateer (probably Tanker 85) passing through smoke over an Oregon fire in a forest of mixed tree types. Firefighters have quantified the best loads for different fuel types, and modern air tanker pilots drop according to the requirements of the particular fire. (Kenn Knackstedt)

Watch That Vortex

The march toward larger air tankers, including converted jumbo jets, exacerbates the potential for dangerous impact along the fire line from wake vortices. Larger aircraft generally make larger vortices, and instructions to fire workers say: "A vortex will feel like a wind gust which lasts from 15 to 20 seconds normally. This may be enough to initiate torching and spotting." As an aircraft passes overhead, its wings generate a pair of spiral vortices that typically sink toward the ground at a rate of 300 to 400 fpm.[249] In still air, with no other noise present, the vortices from a large aircraft may sometimes be heard, sounding like a flag flapping in the wind.

Graceful Davis wing lends aesthetics to this view of a Privateer dropping its payload on a Nevada fire in July 1994. (Keith Conrad)

An over-the-shoulder glance from a hard-hatted firefighter sees Privateer Tanker 123 fighting fire south of Ely, Nevada, in July 1994. (Keith Conrad)

Neptune Tanker 09 passes a wall of smoke as it releases on the Zaca fire in California in 2007. (Darrell Brooks)

A PB4Y-2 releases retardant while banking. Oval cowlings indicate the bomber had not been upgraded to Super Privateer status when the photograph was taken. (USFS)

Privateer air tankers could maneuver surprisingly well; wartime crews in ancestral Liberator variants were known to dogfight the enemy in their four-engine bombers. (USFS)

Ground firefighters, sometimes including loggers, are vital to the final containment of a fire. Their dangers can include freshly logged timber, shorn of limbs and capable of rolling downslope as fire burns away underbrush. Photo taken on Clark Fork fire in northern Idaho, July 1973. (Author Photo)

Calling in an Air Strike

Fire management officials must decide when to use, or stand down, air tanker operations. Guidance to planners published in 1996 includes: "The best and most cost effective use of retardant is clearly initial attack. Most firefighters will also agree the best use of retardant is when there are ground forces available to follow up retardant drops." Issues to be weighed before calling for air tankers on a fire include:

- Danger to life: If firefighters, cooperators, and the public are threatened, order air tankers immediately.
- Property danger: Homes adjacent to public land have the highest priority; structures adjacent to or on public land are the next highest priority; other improvements like fences and recreation sites become the priority if homes or structures are not threatened.

Illuminated wall of smoke provides dramatic yellow gold backdrop to a Beechcraft turboprop lead plane over the Zaca fire in California in summer 2007. (Darrell Brooks)

- Cultural or historical sites, watersheds, range improvements: Use retardant to protect or prevent spread into these areas.
- Threatened and endangered species: Check local policy on use of retardant.
- Timber stands: Does the value of the resource justify the cost of retardant drops?[250]

Air tankers are considered part of a multi-resource plan of attack, and can be used to buy time for ground forces building a fire line. Airdrops can tie together sections of fire line where line building is difficult and slow, and can cool down a section of fire line to assist ground forces. Retardant can also pretreat the combustible foliage in advance of line building. Fire retardant has its limitations—it is ineffective for direct attack when flame length exceeds 8 feet. One tactic of air tanker employment is called herding, where the fire head is directed by placement of retardant. Another tactic, cooling, reduces the intensity of a fire to allow crews or equipment to work in the vicinity.

Planners need to ascertain how many air tankers are available—always a variable number given potential additional fire events and aircraft mechanical problems—and compute flight times to determine if the number of aircraft and the distance involved will permit delivery of an adequate volume of retardant to be effective for a particular blaze. If winds kick up at the fire site, retardant drops generally lose effectiveness when wind speed exceeds 20 mph. Smoke over the fire site can inhibit pilots' abilities to place retardant where needed. Dropping into the wind is said to give the best accuracy. For safety, air tanker pilots favor making run-ins and drops down-sun, downhill, and upwind. "Asking pilots to do otherwise could jeopardize their safety," fire planners are told.[251]

When directing the building of a retardant line, the same criteria used in designing ground fire lines pertains. The retardant line needs an anchor point such as a natural barrier to fire, a road, or a segment of fire line already built by ground crews. Crews are advised to avoid making sharp angles in a retardant line. Direct attack, where retardant is placed on or directly adjacent to a fire has three variations practiced by wildfire teams. Initial first attack is a direct attack across the head of a small fire. A flank attack needs to have an anchor point—a firebreak or natural block to the fire—to keep the blaze from outflanking the retardant line. Spot fires can sometimes be cooled down and corralled by retardant until ground crews arrive. Crews are advised to "use direct attack only when ground support is available or extinguishment is feasible."

Operators may opt for indirect attack methods, laying retardant away from the edge of the fire in anticipation of the fire's migration. Indirect attacks can include pretreatment of fuels to prepare for a deliberate backfire, set by ground crews to limit the spread of a fire in a particular direction. Indirect attacks can also be called to widen or strengthen constructed control lines, natural fuel brakes, or other barriers, to prevent the escape of the fire.

A lime green State of California bird dog Cessna O-2, photographed at Stockton, California, in June 1978, provided coordination over fire sites until replaced by fast OV-10s. (Author Photo)

In choosing the optimum use of air attack, fire managers are advised: "Don't expect miracles from retardant."[252] After more than a half-century of air attack, the lesson is clear: retardant drops are a remarkable tool for speeding fire suppression to remote areas, but the effectiveness of air tankers is dependent upon the availability of ground forces to leverage the work of the bombers.

Fixed-wing air tankers normally orbit in a left-hand pattern at a fire site. Perhaps this is an artifact from the early days of single-engine PT-17 biplanes and TBM torpedo bombers, where propeller torque makes left-hand turns easier; possibly it is the result of the pilot sitting on the left side of the cockpit in larger aircraft, with left banking turns affording a better view of the fire scene below. When ground forces on scene are calling for airdrops on specific spots, they are instructed to try to visualize what the pilot sees from the air overhead, and call out significant landmarks to cue the pilot.

Signal mirrors are viable: "Use signal mirrors—stand in drop location (when safe) for identification and move away before the drop," crews are told. "Use a flag—take bundles of brightly colored flagging, put them on a stick and wave back and forth to get attention of pilots... Describe target from your location and explain mission. The pilot (lead plane or air tanker) will decide drop technique and flight path. Be sure you know aircraft intentions before the drop. Inform and assure pilot when all personnel are out of the drop area. Supervisors have responsibility for the safety of all resources assigned to them."[253]

Ground firefighters are coached on site safety: "Clear the area of the drop—move back in as soon as the aircraft has left the area—take advantage of the retardant. Caution your ground forces to watch their footing when working in the area where the drop has been made, as wet retardant is slick. If the retardant has been dropped across a highway, slow down the traffic or wash it off; it makes cars slip and slide too. If working in timbered areas be alert for snags, treetops, or the possibility of other falling debris knocked loose by retardant. Be cautious of low drop heights by aircraft. The resulting retardant drop will pick up and move rocks, dirt, brush, logs, fire tools, engines, etc. The smaller airborne materials will travel at the drop speed of the retardant."[254]

Bird Dog

Usually where there's smoke there's a bird dog. Air tankers arriving over a fire site get up-to-date situation reports and directions on where to place their retardant loads from a lead plane pilot flying in the bird dog aircraft. Frequently the bird dog leads the air tanker on the run-in, showing where the lead plane pilot wants the retardant released. In the history of aerial firefighting, bird dogs have been everything from a Grumman Goose twin-engine amphibian to Cessna 182s and 210s. The pioneer California biplane air tankers of 1956 used a Piper Tri-Pacer as a bird-dog spotter, with a Forest Service officer onboard directing drops. A tantalizing Alaskan anecdote from 1961 identified the *Pink Lady* (a nicknamed P-51 Mustang), which led Alaska air tankers as a bird dog.[255]

The California Department of Forestry and Fire Protection went shopping for a bird dog fleet in 1974, and obtained 20 Cessna O-2 forward air control (FAC)

Pristine beauty of Sproat Lake on Vancouver Island surrounds Forest Industries Flying Tankers' Grumman Goose used as a bird dog for the Martin Mars tankers when photographed taxiing up the ramp in 1967. (Author Photo)

aircraft from military storage at Davis-Monthan Air Force Base in Arizona. A crew from the California Conservation Corps worked under the supervision of a California forestry battalion chief who was an FAA certificated mechanic with inspection authorization, and together they resurrected their palletized O-2s into a viable state bird dog fleet by 1976. The O-2s served California for two decades. By 1993, the quest for ever-faster, more capable air tankers was mirrored in California's quest for the next-generation bird dog. Sixteen surplus military OV-10A turboprop observation aircraft—the successful outcome of the development in the 1960s of a purpose-built counter-insurgency aircraft with great maneuverability and visibility—were transferred to California that year. Typically 13 are in service as needed over fires in California.[256]

The USFS used a fleet of Beech Baron BE-58P twin-engine general-aviation aircraft as bird dogs (lead planes for large air tankers) and for utility transportation. The Barons have appeared on the news and in publications dramatically leading air tankers over fires. Nineteen USFS Barons were counted when the blue-ribbon panel looked at air tanker operations in 2002. At that time, the panel noted an inherent safety issue in the way the Barons were occasionally employed. In the mountainous western United States, Barons sometimes had to fly at altitudes and air temperatures that put single-engine performance below the aircraft's rated single-engine service ceiling. That is, at the altitudes and air temperatures sometimes faced by the Barons, they would be unable to maintain a rate of climb of 50 fpm if one engine failed. As far back as 1999, an interagency group looked into replacing the Barons with a higher-performing aircraft—one that could maintain a single-engine service ceiling of 10,000 feet at gross allowable weight on especially hot days.[257] Turboprop Beechcraft King Airs now fly lead plane sorties.

The sophistication of bird dog activity led to the development of lead plane tactics, wherein pilots guided air tankers through drop passes, as well as the use of air attack bosses, typically experienced ground firefighters who provided command-and-control functions overhead and acted as observers for the ground-based fire incident commander. Around 2000, the BLM cleaned up the logistics of this by combining lead plane and air attack activity in one aircraft. Subsequently, the USFS adopted this joint overhead aircraft, which flies under the acronym of ASM, or Aerial Supervision Module.

Tanker School

Chico Aviation, part of Aero Union, started operating an off-season school for air tanker pilots in 1972. During fire season, the school's assets—aircraft and pilots—were busy fighting fires. Applications sometimes ran as high as 750, with perhaps only a half-dozen students making it into the program. No place for beginners, the school demanded credentials including a commercial pilot's license, instrument rating, and at least 1,200 flying hours as pilot-in-command. Hotshots who thought they could do anything were quickly sobered by the need for technique, not testosterone, in the profession of air tanker pilot. With 5 to 10 hours of flight time in the curriculum, students practiced retardant drops using cheap water instead of full-up retardant.[258]

Before-and-after study shows a fresh, surplus P2V-7 variant behind a modified P2V-7 air tanker of Aero Union. Canopies bulging upward are a quick clue to P2V-7s versus older P2V-5s also converted to become air tankers. (Donald Keller)

Chapter Five

TANKER TALES: CREW STORIES

Whirling Hamilton Standard prop blades and retardant hoses on the tarmac are reasons enough to pay close attention to air tanker ground operations designed to get tankers in and out of a base quickly, and safely. TBM is seen at Coeur d'Alene, Idaho, in 1973. (Author Photo)

By now the notion of the air tanker pilot as a high-adrenaline, skilled risk-taker is well established. Penn Stohr, who grew up with the corrugated Ford Trimotors of Johnson Flying Service in Missoula, Montana, witnessed and participated in a fair share of the excitement of air tankers. He recalled the 300-gallon Ford Trimotor tanker, converted from a timber spray plane. It had a rudimentary drop gate: "They just cut out some of the corrugations" in the flat belly, Stohr explained. The latch mechanism reminded him of a European beer bottle, with an over-center locking mechanism to hold the gate closed. A cable to the gate was within the crew's reach. "When you pulled it, the door dropped open and stayed open," he recalled. Once the aircraft was back on the ground, loaders reset the door before tanking another 300 gallons of retardant.[259]

Stohr was a college student in Missoula when he began flying for Johnson Flying Service to work his way through school. His father had died in the crash of a Ford Trimotor, and Stohr joined Johnson too late to fly the Ford as an air tanker, but he did use it as a sprayer. Beloved as an airliner and an aviation milestone, the Trimotor may not have been the ideal air tanker, in Stohr's opinion. "It wasn't very good at all," he said, because of its narrow speed range between stall and cruise in the tanker/sprayer mode. Cruising at maybe 90 to 100 mph, the vintage Trimotor was "easily outrun" by other tankers, Stohr explained.[260]

Following early tests of surplus TBM Avenger torpedo bombers in California, Johnson Flying Service bought into the Avenger fleet by 1957-1958, Stohr recalled. "Once they were stripped they were a great

Tanker 22 had a colorful pre-firefighting career, complete with Skyhook inflight personnel pick-up capabilities as showcased in the James Bond movie **Thunderball.** ***Photographed in Evergreen markings in April 1981 at Alamogordo, New Mexico, this view shows what appears to be a non-standard tailwheel.*** (Keith Laird)

B-17G N3509G (44-85778) swapped its blunt Plexiglas nosepiece for a fiberglass streamline bullet shape, as seen at Wenatchee on 23 September 1972. At the time, this Fortress carried a dent in the lower slope of the forward fuselage, sustained in a midair collision with a large raptor, probably an eagle. The pilot and copilot, believing the heavy bird was about to smash through the windscreen, instinctively ducked and leaned inward, knocking heads as their only injuries in the encounter. (Author Photo)

airplane for that sort of thing," he added. Politely laconic in his descriptive speech, Stohr opined: "The TBM was probably as reliable as any early-generation air tanker." Nonetheless, when one smokejumper-trained air tanker pilot experienced an engine failure in a Johnson TBM, he took to his parachute and survived the incident. Back then TBMs were in inexpensive abundance at the Navy's storage yard at Litchfield Park, Arizona. "They were giving them away for almost the fuel in the wings down at Litchfield," Stohr said. Things became even better for the single-engine TBM as newer retardants lowered the weight-per-volume. Nonetheless, the Forest Service moved away from single-engine air tankers by the 1970s, and the current use of turbine-powered single-engine SEATs is a Bureau of Land Management-sponsored undertaking, he explained.[261]

Air tanker operators quickly realized the need for one change to their formerly carrier-based Avengers: On flight decks, the Avengers used a small solid, rubber tail-wheel. They traveled only short distances aboard ship, and this device suited their operations on deck. Not so in the air tanker world, Stohr said, where long taxiing on paved surfaces quickly chewed the solid wheels and spit them out. Pneumatic tailwheels were the only viable choice. Stohr recalled another tailwheel challenge in the air tanker industry years later, after Johnson Flying

Though partially obscured by the smoke of the timber-yard fire it is fighting, the Neptune appears to be the first USFS P2V-5 converted at Medford, Oregon, pressed into unorthodox emergency service before the USFS lettering and civil registration numbers have been added to the airframe. This would place the event somewhere between late 1969 and early 1970. (Kenn Knackstedt)

Service had been bought by Evergreen International. B-17 tailwheels and tires were in short supply, and operators experimented with other units to get by. Later, working with a co-op of B-17 owners, more tires were manufactured to meet the needs of B-17s.[262]

Penn Stohr recalled a pilot fatigue test conducted by a government team who came to town to record heart rate and other physiological data on air tanker pilots. He said the pilots were promised no adverse action would be taken against their pilot medical certificates based on data gleaned in the tests. Given the storied independence of fliers in general, and air tanker pilots in particular, it should come as no surprise that the government testers received heart monitor readouts spiked with bogus high readings induced by batteries, and urinalysis samples that were really beer. Stohr said he read a report on the evaluation that called the pilots "high-spirited," which he took to be tacit acknowledgment of the results received.[263]

In the fire season of 1967, Stohr recalled, Johnson pilot Ray Greene, in an A-26 tanker, suddenly saw a lone tree looming in the smoke of the fire site, at an otherwise cleared and logged patch of ground. Called a seed tree, the lone pine was left standing to help repopulate the forest over time. The pilot, who enjoyed filming his fire sorties with a cockpit-mounted movie camera, could not avoid impact with the seed tree. His movie film subsequently showed the onrushing evergreen, and the abrupt jostling of the A-26 on impact. Deft pilotage brought Greene home with pine needles and kindling mashed into the cylinders of his bomber's shut-down left engine as stark testament to the hazards of the job.[264]

Penn Stohr gained much experience at Missoula: "At Johnson Flying Service... you flew every airplane that was in their inventory," he explained. He instructed in Cessna 150s, carried smokejumpers in a DC-2 and DC-3, performed air tanker sorties in Johnson's only B-25, flew miscellaneous Ford missions, TBM tanker flights, helicopter flights, and by the time he became Johnson's chief pilot in 1974, Stohr was piloting four-engine Lockheed Electras as transports. With Johnson's acquisition by Evergreen, Penn Stohr had the opportunity to fly Evergreen's B-17G and other aircraft in the Evergreen fleet, retiring in 2002, but returning to test Evergreen's first 747 Supertanker, a convertible cargo 747-200 variant. Subsequently, Evergreen evolved the Supertanker as a dedicated 747-100 airframe with permanent air tanker equipment installed, he explained.[265]

Cal Butler, the namesake and icon of Butler Aircraft in Oregon for many years, recalled when he first approached the government to get his newly acquired B-17G approved for USFS use, and to get his own FAA B-17 Type Rating as a pilot in about 1962. Skilled as a flier, Butler quickly embraced the B-17 on the ferry flight to have tanks installed. He still needed to

The most glamorous Privateer air tanker paint job undoubtedly belonged to Tanker 121 (N2871G), photographed at Fox Field, Lancaster, California, circa 1997. (Author Photo)

cross the hurdle of getting his tanker rating by demonstrating a drop in the bomber to an FAA observer in Medford, Oregon. Butler said his plan was to carry only about 1,500 gallons for that inaugural drop just to be on the safe side. But the person filling the B-17's new tanks topped them off until they overflowed, indicating a load of 2,470 gallons. Butler faced a dilemma. He couldn't drain any water from the tanks without arousing curiosity from the FAA and the Forest Service, so he stoically proceeded to take off with a full load.

His previous experience in single-engine TBM tankers included drops with no flaps, so he set up the big B-17's approach this way over the Medford field. Cruising from east to west at about 100 feet, Butler salvoed all four doors and felt the B-17 jump about 50 feet in altitude—or at least it felt that way in a heart-stopping moment. He said the nose pitched up precariously due to the lack of 15 or 20 degrees of flaps that were needed to keep it level. Butler and copilot Bob Nelson held the control yokes full forward and used full throttle to regain normal flying attitude. Decades later, this B-17, Tanker 65, was restored and flown to England for permanent display.[266]

There are other air tanker pilot stories that swirl and linger in the air like trails of retardant at the end of a good drop. Whether you take them as truth or take them as heroic legend, they nonetheless add to the lore about that special breed of air tanker pilots. A no-nonsense air tanker pilot is said to have expressed concern to a mechanic about the condition of one of his bomber's main tires during a day of multiple sorties. Breaking for lunch, the pilot asked the mechanic to replace the tire in question, but the mechanic, so the story goes, opined that the tire was fine as is. The pilot had more riding on this than did the ground-bound mechanic. The pilot quietly returned to his aircraft and extracted a handgun he owned, and put one round through the tire. He walked off for lunch, satisfied a newly mounted tire would await him upon his return to duty.

B-17 Tanker 42 (replacing a Privateer that formerly held this number) carried a dent in the aluminum on the lower slope of the nose in 1972, said to be the unhappy result of a midair with an eagle. Hazards on fire runs were more than just other aircraft; birds and even trees got in the way sometimes. One of the pilots confirmed the mishap, saying the pilots glimpsed the moving bird too late to avoid it.

Though typically regarded as protectors of wildlands, air tankers have dropped over populated regions at times to bring their services to bear on urban fires. A lumberyard fire in Oregon was treated by drops from the first P2V tanker, as captured on film by photographer Kenn Knackstedt.

And just in case this story is still circulating, it is worth noting the Web site www.snopes.com, among others, debunks tales of a diver (or a fisherman) being crammed into the scoop of an amphibious air tanker on a water pickup run, only to be discharged over a fire site. Really now…

The air tanker profession has evolved since the days of Stearmans, and Ford Trimotors, and TBMs, and will continue to do so. Nothing lasts forever. Except memories and memorials. Let this volume contribute to those.

Score: A-26s, 2; trees, 0—pine needles seen packed around the cylinders of this A-26's port engine tell a potentially foreboding tale that had a happy ending. (USFS)

That's a fresh pine remnant held aloft in front of the A-26 wing that carried it back from a fire site. The Johnson Flying Service pilot was racing though smoky skies over a clearcut area. Only one tree remained—a seed tree left standing to help propagate a new round of timber to be harvested. The lone pine was seen too late to avoid, and the bomber took a direct hit. The pilot made it home safely on the other engine. (Penn Stohr)

Temporarily out of the fight, B-17 Tanker A18 of Black Hills Aviation undergoes engine work while A-26 Tanker A17 lands at Coeur d'Alene during fires in Idaho in the summer of 1973. (Author photo)

TANKER CALL SIGNS AND REGISTRATIONS

Air tankers in the United States have been identified by call signs for decades. To the extent possible, the following roster lists call signs assigned to particular aircraft, with civil registration numbers included where known.

Some call signs are numeric only; the subsequent convention of alphanumeric call signs results in some tankers flying under different call signs at different times in their careers. The assignment of the same call sign to different aircraft shows up over time. And on a few occasions, the same civil registration N-number has been reassigned to the same kind of aircraft after the loss of the first aircraft bearing this number. (PB4Y-2 Privateer registration number N6884C was assigned first to aircraft BuNo. 66284 which crashed in 1959, and subsequently to Privateer BuNo. 59701.) A change in nomenclature identifying navy aircraft sometimes results in air tankers being identified in multiple ways. Lockheed Neptunes show up generically as P2V or P-2; model variants can lead to some further identified as P2V-5, P2V-7, or SP-2H. C-54 Skymasters and their DC-4 commercial counterparts likewise show up interchangeably on some lists, to the possible chagrin of strict roster keepers.

Changes in tanker numbers, changes in ownership, and variations inherent in the many sources available on air tanker numbers may allow errors to creep into any attempt at making a comprehensive air tanker roster. This appendix represents an earnest basis that is subject to continual review and refinement. (Abbreviations: Aviation = Av; Service = Svc; U.S. Forest Service = USFS)

Call Sign	*Civil Registration*	*Aircraft*	*Comments*
B-17			
E84, C84, C44, 04	N17W	B-17F	Max Biegert, Aircraft Specialties, Globe Air, Museum of Flight
35, 34, A34, E34	N1340N	B-17F	Aero Enterprises, Aero-Flite; turboprop mod, 1970 crash
E85, A10, 10	N66573	B-17G	Ewing Aviation, Black Hills Aviation; crashed 1979
C64, 64	N621L	B-17G	Aircraft Specialties; crashed 1975
E16, 42, 102	N3509G	B-17G	Leo Demers, Aero Union, Butler, Central Air Service
C12	N6694C	B-17G	Ewing Av, Black Hills Av; former Curtiss-Wright, 1980 crash
30, 37	N9323R	B-17G	Black Hills Av, Aero-Flite, Kinney Air Tankers
C34, 34	N3193G	B-17G	Aircraft Specialties, Globe Air
C54	N620L	B-17G	Aircraft Specialties; crashed 1973
E61, 61	N3702G	B-17G	E. A. Neely, TBM Inc., Aero Union
E17, 17	N9323Z	B-17G	Aero Union; later CAF *Sentimental Journey*
E18, 18	N9324Z	B-17G	Aero Union; crashed 1971
75, E78, E68, 68	N3703G	B-17G	Fast Way Air Service, TBM Inc., David Tallichet
C24, 89	N9563Z	B-17G	Aircraft Specialties, Globe Air
99	N93012	B-17G	Aircraft Specialties, Globe Air
C71, 71, 22	N809Z/N207EV	B-17G	Intermountain Aviation, Evergreen Aviation
A18, 09	N66571	B-17G	Black Hills Aviation; National Air and Space Museum
F71, 71, D1, 18	N5233V	B-17G	Rogue Flying Services, Aero Ag, Aero Union
B11, E56	N73648	B-17G	Cal-Nat Airways, Black Hills Av; crashed July 1972
15, 65	N5237V	B-17G	Cal Butler, Butler Aircraft, TBM Inc.
C19, 19	N5230V	B-17G	Aero Union
	N3678G	B-17G	Crashed October 1962 in Nevada
C-97			
67	N9538C	YC-97	Flew in 1961 season
84, 97	N1365N	C-97G	Hawkins and Powers
747			
947	N470EV	B-747-200	Evergreen
979	N479EV	B-747-100	Evergreen

Call Sign	*Civil Registration*	*Aircraft*	*Comments*
AT-11/C45			
E2	N6591C	AT-11	Willows Flying Svc
E16	N66666	AT-11	Willows Flying Svc
E24	N68Q	AT-11	P&B Aviation
E25	N79A (?)	AT-11	Aero Atlas, Red Bluff
E26	N6961C	AT-11	Aero Atlas, Red Bluff
E28	N6956C	AT-11	P&B Av, Red Bluff
	N64970	C-45	State of Florida, early 1960s
	N9056Z	C-45	Possibly State of Nevada
CL-215			
215	N215NC	CL-215	State of North Carolina
262	N262NR	CL-215	Aero-Flite
263	N263NR	CL-215	State of Minnesota
264	N264V	CL-215	Aero-Flite
266	N266NR	CL-215	State of Minnesota
267	N267V	CL-215	Aero-Flite
PBY			
E48	N9752Z	PBY-5A	
49	N2886D	PBY-6A	Sis-Q Flying Service, SLAFCO
E93	N2887D	PBY-6A	
E54, 54, 85	N6453C	PBY-6A	Cal-Nat Airways, Hemet Valley Flying Svc
F47	N6455C	PBY-6A	Liston Aircraft
E38, 84	N6456C	PBY-6A	Sonora Flying Svc
E49	N6457C	PBY-6A	Sonora Flying Svc
F46, E33, 83, 84	N6458C	PBY-6A	Liston Aircraft, Hemet Valley Flying Svc
F11	N6459C	PBY-6A	Liston Aircraft, Klamath Falls
E87	N2846D	PBY-6A	Jadon Enterprises
	N6986C	PBY-6A	Crashed June 1961, Winslow, AZ
E94	N7082C	PBY-6A	Routh Aircraft
53		PBY	SLAFCO
98	N31235	PBY	SLAFCO
158	N98252	PBY	SLAFCO
285, 85	N85U	PBY	Flying Fireman
12		PBY-6A	Aurora Air; Ace Demers; Air Ag (may be Tanker E54, above)
PB4Y-2			
C50	N7237C	PB4Y-2	Crashed 27 June 1974
C30, 30	N3739G	PB4Y-2	Flight Enterprises, SST Inc., Douglas County Av, T&G Av
A25, B25, 125	N6813D	PB4Y-2	Hawkins and Powers; ditched 1975
F42, 42	N6816D	PB4Y-2	Wenairco, Dempsey; destroyed by fire, 1972
B26, 126	N7962C	PB4Y-2	Hawkins and Powers
F84, 84, 127	N6884C	PB4Y-2	Rosenbalm Av, Hawkins and Powers (Two PB4Y-2s used this registration after loss of first aircraft)
A20, B21, 121	N2871G	PB4Y-2	Hawkins and Powers
B22, 122	N2870G	PB4Y-2	Avery Aviation, Hawkins and Powers; dismantled after mishap
B23, 123	N7620C	PB4Y-2	Hawkins and Powers; crashed 2002
124	N2872G	PB4Y-2	Hawkins and Powers
	N7974A	PB4Y-2	Crashed July 1968 near McGrath, AK
F85, 85	N3191G	PB4Y-2	Rosenbalm Av, ditched in Diamond Lake, OR; scrapped
A-26			
26	N9402Z	A-26C	Idaho Air Tankers, Reeder Flying Svc
01	N4060A	A-26B	Kem Air, Lynch Air Tankers; crashed 1983
58	N4805E	A-26B	Lynch Air Tankers
F20, A20, 16	N5457V	A-26B	Hillcrest, Wilson Av Industries, Butler Aircraft
F74, 74, 107	N74874	A-26B	Kreitzberg Av, Rosenbalm Av, W. Dempsey, Bruce Kinney
C29	N4820E	A-26C	Flight Enterprises
C55	N8026E	A-26C	Kreitzberg Av, Aero-Flite, Aero Union, W. Dempsey
B27, 59, A28	N3427G	A-26C	Kern Air, Lynch Air Tankers; crashed 1976

Call Sign	Civil Registration	Aircraft	Comments
A17, 56	N3426G	A-26C	Johnson Flying Svc, Evergreen, Lynch
F56, 56	N86469	A-26C	Aero-Flite, Troutdale, OR
F29, 108	N3428G	A-26C	Rosenbalm Av, Dempsey, Kinney, Custom Farm, Central Air Svc
F12, 12, 104	N5544V	A-26C	Rosenbalm Av, Dempsey, Kinney
C11	N9996Z	A-26C	Donaire Inc., G. Stell
F77, 57	N9425Z	A-26C	Central Oregon Aerial Co., Lynch Air Tankers; crashed
F83, 83, 106	N7954C	A-26C	Rosenbalm Av, Flick Aviation, Dempsey, Central Air Svc
F57	N4819E	A-26B	Rosenbalm Av
27	N9403Z	A-26C	Reeder Flying Svc
C32	N4822E	A-26C	Unknown
C32	N8017E	A-26	
28	N91354	A-26B	Reeder Flying Svc
F23, 23	N9161Z	A-26C	Hillcrest Aircraft, Butler Aircraft
59	N4818E	A-26C	Lynch; see also other tanker B27/59, above
F24	N9159Z	A-26C	Hillcrest
F76	N9423Z	A-26B	Central Oregon Aerial Co.
C12	N9301R	A-26C	Aircraft Specialties
C13	N8037E	A-26C	Moseley Av
C31	N8629E	A-26B	Flight Enterprises
C33	N9401Z	A-26C	Donaire Inc.
55	N74883	A-26	Lynch Air Tankers
6G		A-26	
B-18			
120	N52056	B-18	Hawkins and Powers
	N66267	B-18B	Roberts Aircraft, Boise
C-54/DC-4			
153, 161	N82FA	C-54G	ARDCO
18, 76, 16	N76AU	C-54Q	Bicentennial paint in 1976
65, 165	N8502R	C-54	Spiva, WAIG, TBM Inc.
119	N406WA	C-54	WAIG Aircraft Inc., ARDCO
151	N460WA	C-54E	WAIG Aircraft Inc., ARDCO
152	N9015Q	C-54D	ARDCO
160	N96358	C-54E	Aero-Flite Inc.
166		DC-4	
14	N62297	C-54	Aero Union
109	N68160		
111	N96451	DC-4	Central Air Svc
15	N2742G	C-54	Aero Union
147		C-54/R5D	BuNo. 90392
147	N67040	DC-4	Central Air Svc
02	N11712	C-54	Aero Union
18	N4218S	C-54	Aero Union
13	N62342	DC-4	Aero Union
16	N3303F	DC-4	Aero Union
19	N3373F	C-54E	Aero Union
19	N4989P	C-54G	Aero Union
113	N99212	DC-4	Douglas County Av
117	N31356	DC-4	California National Air Svc, Central Air Svc
146	N67061	DC-4	Central Air Svc
148	N67062	DC-4	Central Air Svc
149	N67041	DC-4	Central Air Svc
150	N67034	DC-4	Central Air Svc
154	N98AS	DC-4	Globe Air
155	N67017	DC-4	Globe Air
156	N67018	DC-4	Globe Air
157		DC-4	Ken Spiva
162	N3054V	C-54G	Aero-Flite
26	N76AU	C-54	Aero Union

Call Sign	Civil Registration	Aircraft	Comments
DC-6			
20	N80MA	DC-6	Sis-Q Flying Svc
21	N90MA	DC-6	Sis-Q Flying Svc
44	N444SQ	DC-6B	Sis-Q Flying Svc
45	N555SQ	DC-6B	Sis-Q Flying Svc
46	N111AN	DC-6B	Sis-Q Flying Svc
47	N666SQ	DC-6	Sis-Q Flying Svc
48	N888SQ	DC-6B	Sis-Q Flying Svc
51	N999SQ	DC-6B	Sis-Q Flying Svc
52	N66DG	DC-6B	Aero Retardant Inc.
54	N333SQ	DC-6B	Sis-Q Flying Svc
97, 68	N90739	DC-6	TBM Inc.
82	N37515	DC-6	Rosenbalm Av
84	N37565	DC-6B	Rosenbalm Av
85	N37585	DC-6B	Rosenbalm Av
66	N6353C	DC-6	TBM Inc.
DC-7			
110	N823D	DC-7B	Bruce Kinney
67, 167	N6318C	DC-7	TBM Inc.
33	N4887C	DC-7B	Douglas County AvInc., Int'l Air Response
60	N838	DC-7B	TBM Inc.
61	N848D	DC-7B	TBM Inc.
62	N401US	DC-7	TBM Inc.
29	N4889C	DC-7B	Douglas County Av Inc.
35	N5903	DC-7C	Douglas County AvInc.
31	N51701	DC-7B	Douglas County AvInc.
66	N6353C	DC-7	TBM Inc.
115	N90802	DC-7C	Douglas County AvInc.
28	N90804	DC-7C	Douglas County AvInc.
32	N9734Z	DC-7C	Douglas County AvInc.
DC-10			
910	N450AX	DC-10-10	10 Tanker STC, LLC
C-119			
140	N807Z	C-119	Hawkins and Powers
133	N383S	C-119	Aero Union, Hemet Valley Flying Svc
137	N3559	C-119F	To AMC Museum at Dover AFB
140	N3560	C-119F	Hawkins and Powers
29, 139	N3935	C-119	Hawkins and Powers
136	N5216R	C-119F	Hawkins and Powers
138	N8092	C-119F	Hawkins and Powers; crashed 1981
140	N8093	C-119	Hawkins and Powers
136	N8094	C-119F	Hawkins and Powers
28, 138	N8682	C-119	Aero Union, Hawkins and Powers
134	N8832	C-119F	Hawkins and Powers
88	N13742	C-119J	Hemet Valley Flying Svc; crashed
E12, 81	N13743	C-119C	Aero Union, Hemet Valley Flying Svc
13, 86	N13744	C-119	Aero Union, Hemet Valley Flying Svc
14, 82	N13745	C-119	Aero Union, Hemet Valley Flying Svc
87	N13746	C-119	Hemet Valley Flying Svc
36	N15509	C-119	Douglas County Av Inc., Hawkins and Powers
135	N48076	C-119	Hawkins and Powers; crashed September 1987
Fairchild C-123			
63	N3142D	C-123	Aero Union
Ford Trimotor			
2F	N9612	4-AT-E	
	N8407	4-AT-E	

Call Sign	Civil Registration	Aircraft	Comments
AF-2			
	N3143G	AF-2S	
E30, 30	N3144G	AF-2S	Aero Union
E21	N9995Z	AF-2S	Aero Union
TBM			
	N104Z	TBM-3	USFS
	N109Z	TBM-3	Crashed August 1960
A13	N7922A	TBF-1C	Photographed at Missoula, MT, 1965
?	N9394H	TBM-1C	Paul Mantz Air Svc; TBM air tanker testbed, 1954
D6, A13, 13	N9010C	TBM-3	Hillcrest Av, Johnson Flying Svc; see other A13, above
D19	N9592C	TBM-3	Aerial Applicators
E26	N9078Z	TBM-3	Daro Inc., Sis-Q Flying Svc
E25	N68663	TBM-3	Klamath Aircraft Inc., Sis-Q Flying Svc
D13, 13	N7030C	TBM-3E	Idaho Air Tankers, Reeder Flying Svc; see other 13, above
E63, E28	N9083Z	TBM-3	Riverside Aircraft, Aero Union; crashed 1973
?	N6822C/N5567A	TBM-3E	TBM Inc.
D20	N8398H	TBM-3	Aerial Applicators
E46, 33	N9711Z	TBM-3E	Sierra Av, Wen Inc., P&B Av, Sis-Q Flying Svc
B18	N103Z/N10361	TBM-3	USFS circa 1956, then tanker B18 Seeley Flying Svc, Air Tankers Inc.
D1, 1	N9590C	TBM-3	Idaho Aircraft, Boise; Dennis G. Smilanich
A21	N3969A	TBM-3E	Edgar L. Thorsrud, Johnson Flying Svc
E79	N9710Z	TBM-3	Columbia Flying Svc, George C. Abell
55	N9591C	TBM-3	Sonora Aviation
D16	N7029C	TBM-3	Aerial Applicators
C96, E96	N9679C	TBM-3	Routh Aircraft, Hemet Valley Flying Svc
92, 24	N7961C	TBM-3	Daro Inc., Sis-Q Flying Svc
23	N73642	TBM-3	Reeder Flying Svc
A6	N9597C	TBM-3	Cal Butler, Johnson Flying Svc
E66	N1366N	TBM-3	TBM Inc., possibly leased to Aero Union; crashed 1973
E36, E27	N1369N	TBM-3	TBM Inc., Sis-Q Flying Svc
A12	N6824C	TBM-3E	Wilson Aviation Industries, Johnson Flying Svc
1, A11	N7014C	TBM-3E	Johnson Flying Svc
75	N3356G	TBM-3	Hemet Valley Flying Svc; crashed 1970
A7	N7015C	TBM-3E	Johnson Flying Svc
C3, E39, C39	N9927Z	TBM-3E	Aircraft Specialties, Globe Air
37	N6825C	TBM-3E	Ralph M. Ponte, Cal-Nat Airways, Hemet Valley Flying Svc; crashed 1971
E80	N9547Z	TBM-3E	Aero Crop Corp., Reeder Flying Svc
C97, E97	N7161C	TBM-3E	Routh Aircraft, Hemet Valley Flying Svc
E52	N9434Z	TBM-3	Sonora Flying Svc, Fire Flyers Inc., Hemet Valley Flying Svc
E95	N9860C	TBM-3	Routh Aircraft, Hemet Valley Flying Svc; crashed 1970
E71	N7229C	TBM-3E	Hemet Valley Flying Svc; crashed 1971
C50	N7219C	TBM-3E	
E58	N6827C	TBM-3E	TBM Inc., Hillcrest Aircraft
A14	N3249G	TBM-3	Johnson Flying Svc
D17	N7858C	TBM-3	Aerial Applicators
D15	N7848C	TBM-3	Aerial Applicators
E74	N7833C	TBM-3	Hemet Valley Flying Svc
A15	N3251G	TBM-3	Johnson Flying Svc
D10, 10	N4170A	TBM-3E	Reeder Flying Svc, Twin Falls
D11, 11	N4171A	TBM-3E	Reeder Flying Svc
D8, 8	N4168A	TBM-3	Reeder Flying Svc
C56, 56	N9433Z	TBM-3E	Sonora Flying Svc, Sonora Aviation, Capitol Air Inc.
E76	N9548Z	TBM-3E	Rector Air Svc, Hemet Valley Flying Svc
E51, 57, 12, D12	N9429Z	TBM-3E	Sonora Flying Svc, Reeder Flying Svc
E81, C25	N9590Z	TBM-3E	Cisco Aircraft, Aircraft Specialties, Globe Air
9	N4169A	TBM-3E	Reeder Flying Svc
D18	N4172A	TBM-3E	Aerial Applicators
E68	N5168V	TBM	Hemet Valley Flying Svc

Call Sign	Civil Registration	Aircraft	Comments
E74	N7839C	TBM	Hemet Valley Flying Svc
E-19	N9596C	TBM-3	P&B Aviation
E29	N9318Z	TBM-3	P&B Aviation
E38	N9926Z	TBM-3	Aircraft Specialties
E44	N7226C	TBM-3	Sierra Aviation
E45	N6823C	TBM-3	Sierra Aviation, Porterville
E57	N9651C	TBM-3	Cal-Nat Airways
E72	N3357G	TBM-3	Hemet Valley Flying Svc
E73	N9082Z	TBM-3	Hemet Valley Flying Svc
E75	N3356G	TBM-3	Hemet Valley Flying Svc
E89	N9187Z	TBM-3	Columbia Flying Svc
E98	N9708Z	TBM-3	Daro Inc.
E22	N6831C	TBM-3	C. T. Jensen, Sacramento
C34, E56	N9564Z	TBM-3	Aircraft Specialties, Cisco Av
E64	N7239C	TBM-3	Riverside Av
E65	N5260V	TBM-3	Riverside Av
E67	N9595C	TBM-3	Cisco Aircraft, Lancaster
E70	N5170V	TBM-3	Hemet Valley Flying Svc
E82	N9307Z	TBM-3	Cisco Av, Lancaster
	N7157C	TBM-3	D. Goodman, Missoula
D5	N9593C	TBM-3	Clayton Curtis, Boise
D3	N5635N	TBM-1C	Idaho Av, Boise
D4	N7026C	TBM-3	Idaho Av
D7	N7028C	TBM-3	Hillcrest Aviation
C46	N3968A	TBM-3	
C47	N6447C	TBM-3	Sonora Aviation
C53	N6826C	TBM-3	
C55	N6894C	TBM-3	Sonora Aviation
	N9830C	TBM-3	Crashed July 1959
	N9319Z	TBM-3	Engine failure and fire June 1961
	N7032C	TBM-3	Off runway July 1961 Idaho City
	N3250G	TBM-3	Bailout September 1961
	N9428Z	TBM-3	Crashed June 1961
60E	N1366N	TBM-3	Aero Union
37	N6825C	TBM-3	Cal-Nat Airways
55	N7075C	TBM-3	BuNo. 53786
28	N9083Z	TBM-3	Aero Union
83	N9650C	TBM-	BuNo. 69374; Bill Woods
	N108Z; N108Q	TBM-3U	Georgia Forestry Commission
E92, 24		TBM	Daro Inc.; Sis-Q Flying Svc
F7F			
F18, 18, E63	N7654C	F7F-3	Butler, then TBM Inc.; to Naval Aviation Museum, Pensacola
E62, E41	N7629C	F7F-3N	Cal-Nat Airways, then Sis-Q Flying Svc; to USMC Museum, Quantico
F16, 16, E62	N6129C	F7F-3N	Johnson Flying Svc, Butler, TBM Inc.
E31, E41	N6177C	F7F-3	S.V. Flying Svc, Sis-Q Flying Svc; crashed 1974, Ukiah, CA
E42	N7626C	F7F-3	Cal-Nat Airways, Sis-Q Flying Svc
?	N7627C	F7F-3	Cal-Nat Airways, Sis-Q Flying Svc (as hulk)
E59	N7628C	F7F-3	Cal-Nat Airways, Sis-Q Flying Svc (as hulk)
F17, 17, E64, 64	N7235C	F7F-3P	Butler, TBM Inc.
E43	N6178C	F7F-3	Cal-Nat Airways, Sis-Q Flying Svc
E43, E23	N6179C	F7F-3	Fred J. Arnberg, Sis-Q Flying Svc; crashed 1974, Rohnerville, CA
?	N800RW	F7F-3	TBM Inc.
E41, E22	N7238C	F7F-3	Dick Gordon, Sis-Q Flying Svc; crashed 1974, Rohnerville, CA
E32, E40	N7195C	F7F-3	George Kreitzberg, Sis-Q Flying Svc
	N7625C	F7F-3	Crashed September 1962
J2F			
07	N67790	J2F-6	Grumman-designed Duck biplane amphibian

Call Sign	Civil Registration	Aircraft	Comments
S-2			
154	N736MA	S-2	Marsh Aviation
155	N746MA	S-2T	Marsh Aviation
	N402DF		
71	N403DF	S-2	California Dept of Forestry
72	N405DF	S-2	California Dept of Forestry
73	N406DF	S-2T	California Dept of Forestry
70	N400DF	S-2	California Dept of Forestry/Volpar Modification
74	N411DF	S-2T	California Dept of Forestry
78	N412DF	S-2A	California Dept of Forestry/Aero Union
77	N415DF	S-2	California Dept of Forestry
96	N416DF	S-2	California Dept of Forestry
76	N417DF	S-2T	California Dept of Forestry
75	N420DF	S-2T	California Dept of Forestry
70	N427DF	S-2T	California Dept of Forestry
71	N432DF	S-2T	California Dept of Forestry
72	N435DF	S-2T	California Dept of Forestry
100	N436DF	S-2	California Dept of Forestry
73	N437DF	S-2T	California Dept of Forestry
74	N439DF	S-2T	California Dept of Forestry
75, 79	N444DF	S-2T	California Dept of Forestry
76	N436DF	S-2T	California Dept of Forestry
78	N431DF	S-2T	California Dept of Forestry
80	N445DF	S-2T	California Dept of Forestry
94	N446DF	S-2	California Dept of Forestry
93, 81	N447DF	S-2	California Dept of Forestry
81	N449DF	S-2T	California Dept of Forestry
90, 87	N450DF	S-2	California Dept of Forestry
92	N451DF	S-2	California Dept of Forestry
91	N453DF	S-2	California Dept of Forestry
82	N422DF	S-2T	California Dept of Forestry
83	N424DF	S-2T	California Dept of Forestry
100, 85	N438DF	S-2, S-2T	California Dept of Forestry
86	N433DF	S-2T	California Dept of Forestry
88	N426DF	S-2T	California Dept of Forestry
89	N425DF	S-2T	California Dept of Forestry
90	N434DF	S-2T	California Dept of Forestry
91, 70	N428DF	S-2T	California Dept of Forestry
93	N450DF	S-2T	California Dept of Forestry
70, 94, 92	N442DF	S-2, S-2T	California Dept of Forestry
95	N448DF	S-2T	California Dept of Forestry
96	N440DF	S-2T	California Dept of Forestry
100, 71	N441DF	S-2T	California Dept of Forestry
72	N443DF	S-2T	California Dept of Forestry
77	N423DF	S-2T	California Dept of Forestry
80	N404DF	S-2T	California Dept of Forestry
86	N453DF	S-2T	California Dept of Forestry
Grumman Ag-Cat			
C-130			
30	N662U	C-130	Ag Cat
31	N117TG	C-130A	Int'l Air Response
32	N118TG	C-130A	Int'l Air Response
67	N531BA	C-130	TBM Inc.
81	N131FF	C-130A	Hemet Valley Flying Svc
83		C-130	
88		C-130	
130	N130HP	C-130A	Hawkins and Powers
131	N131HP	C-130A	Hawkins and Powers
133	N133HP	C-130	Hawkins and Powers
64	N466TM	C-130A	TBM Inc.
63	N473TM	C-130A	TBM Inc.

Call Sign	Civil Registration	Aircraft	Comments
PV-2			
29	N83L	PV-2	
103	N6651D	PV-2	Bruce Kinney
F37, 37	N6657D	PV-2	Wenairco, Wenatchee, WA
62	N6857	PV-2	
F62, 43, 112	N7086C	PV-2	B&B Flying Svc, Douglas County Av Inc.
F40, 40	N6651D	PV-2	Wenairco, Wenatchee, WA
F30	N7459C	PV-2	Aero Ag, Medford, OR
38	N6856C	PV-2	Douglas County Av Inc.
39	N7080C	PV-2	Douglas County Av Inc.
	N7273C	PV-2	
P2V			
	N126Z	P2V-5	USFS; first Neptune conversion
141	N202EV	P2V-5	Evergreen
142	N203EV	P2V-5	Evergreen
143	N204EV	P2V-5	Evergreen
144	N205EV	P2V-5	Evergreen
145	N206EV	P2V-5	Evergreen
01	N701AU	SP-2H	Aero Union
03		SP-2H	
16	N716AU	SP-2H	Aero Union
18	N718AU	SP-2H	Aero Union
04	N918AP	P2V-7	Black Hills Aviation
09	N4235T	P2V-7	Neptune
139		P2V-7	
140	N140HP	P2V-7	Hawkins and Powers
55		P2V-7	
99		P2V-7	
15	N14448	P2V	Consumed by fire, Wenatchee, WA, circa 1976
12	N96264	P2V-5	Black Hills Aviation, Neptune
55	N355MA	P2V-7	Minden Air
42	N442NA	P2V-7	Neptune
48	N4692A	P2V-7	Minden Air
45	N445NA	P2V-7	Neptune
10	N4235N	P2V-7	Neptune
44	N1386C	P2V-5F	Neptune
07	N1386K	P2V-5	Neptune
07	N807NA	P2V-5F	Neptune
11	N14447	P2V-7	Black Hills Aviation, Neptune
05	N96278	P2V-5F	Black Hills Aviation, Neptune
43	N443NA	P2V-7	Neptune
06	N9855F	P2V-5F	Black Hills Aviation, Neptune
07	N96271	P2V-5	Black Hills Aviation
08	N65170	P2V-7	Black Hills Aviation
P-3			
00	N900AU	P-3A	Aero Union
17	N917AU	P-3S	Aero Union
20	N920AU	P-3A	Aero Union
21	N921AU	P-3A	Aero Union
22	N922AU	P-3A	Aero Union
23	N923AU	P-3A	Aero Union
25	N925AU	P-3A	Aero Union
26		P-3	
27	N927AU	P-3A	Aero Union, Douglas County Av Svc
N3N-3			
E3	N44959	N3N-3	
E4	N45163	N3N-3	
E5	N45233	N3N-3	
E6	N45032	N3N-3	
E7	N45292	N3N-3	

Call Sign	Civil Registration	Aircraft	Comments
E8	N45014	N3N-3	
E9	N45177	N3N-3	
E10	N7752C	N3N-3	
E11	N45251	N3N-3	
E12	N7753C	N3N-3	
E14	N44972	N3N-3	
E15	N45089	N3N-3	
E20	N45057	N3N-3	
E21	N45228	N3N-3	
E23/11	N45084	N3N-3	
E27	N45188	N3N-3	
E47	N45047	N3N-3	
AJ			
E88, 88	N9142Z	AJ-1	AJ Air Tankers, Van Nuys
E77, 77	N9143Z	AJ-1	AJ Air Tankers, Van Nuys
AT-6			
	N188C	AT-6G	Crashed August 1961
	N62701	AT-6B	Crashed September 1961
B-25			
	N88972	B-25D	Alaska air tanker
E30	N9865C	B-25J	Aero Union
32	N9079Z	B-25J	
33	N9655C	B-25J	
C34	N6123C	B-25J	Abe's Aerial Svc, Aircraft Specialties
C35	N9117Z	B-25J	Abe's Aerial Svc, Aircraft Specialties
E91	N10564	B-25J	USFS test aircraft
	N3503G	B-25J	Crashed July 1960
	N9613C	B-25J	Crashed
	N9623C	B-25J	
7	N8163H	B-25J	Alaska air tanker
	N8196H	B-25J	Alaska air tanker
	N9116Z	B-25J	Idaho Aircraft, Boise
	N3337G	B-25J	
	N9080Z	B-25J	Donaire Inc.
	N9613C	B-25J	Crashed July 1960
81	N5256V	B-25J	
	N9856C	B-25J	Idaho Aircraft, Boise
82	N9455Z	B-25J	Paul Mantz Air Svc
4	N7946C	B-25N	Red Dodge Aviation
39	N9877C	B-25J	
P-61/F-15			
	N30020	P-61B	Ranchers Inc.; crashed August 1963
E35	N9768Z	F-15	Aero Enterprises Inc.; crashed Hollister, CA, September 1968
Air Tractor 802			
181	N91092	AT-802A/F	Queen Bee
182	N5035K	AT-802F	Queen Bee
183	N9002K	AT-802F	Queen Bee
184	N9135F	AT-802F	Queen Bee
190	N6159F	AT-802F	Aerospray
250	N6195F	AT-802F	Queen Bee
401	N325BM	AT-802	Fletcher Flying Svc
402	N106LA	AT-802	M&M Air
403	N4120R	AT-802	M&M Air
404	N5202U	AT-602	Western Pilot Svc
405	N319BM	AT-802	Evergreen

Call Sign	*Civil Registration*	*Aircraft*	*Comments*
406	N406BM	AT-802	Evergreen
407	N8523H	AT-802	New Frontier
408	N196PK	AT-502	Sun Valley Dusters
409	N6131K	AT-802	M&M Air
410	N187LA	AT-802	Western Pilot
417	N417BM	AT-802	Evergreen
419	N419FF	AT-802	Fletcher Flying Service
420	N196LA	AT-802AH	Western Pilot
421	N207LA	AT-802AH	Western Pilot
422	N41119	AT-802	Western Pilot
423	N1539X	AT-502	M&M Air
424	N1558W	AT-802	M&M Air
425	N802MA	AT-802	Minuteman Av.
426	N627JP	AT-802	Henry's Flying Svc
428	N428AE	AT-802	G.B. Aerial
429	N8524H	AT-802	Henry's Aerial
430	N802SG	AT-802AH	Western Pilot
434	N5096Y	AT-602	MJ Aviation
435	N602AA	AT-602	Aero Seat; aircraft destroyed
436	N602BG	AT-602	Aero Seat
440	N427NS	AT-802	Queen Bee
441	N8510M	AT-802	M&M Air
444	N90802	AT-802	Queen Bee
445	N9190G	AT-802	Sly's Aircraft Svc
446	N8507N	AT-802	M&M Air
447	N457AT	AT-802	Aerial Timber
448	N157LA	AT-802	Aerial Timber
449	N6162N	AT-802	Aerial Timber
450	N9103A	AT-802	Minuteman
452	N189LA	AT-802	Minuteman
454	N804MA	AT-802	Minuteman
456	N6165U	AT-502	Minuteman
457	N807MA	AT-802	Minuteman
458	N806MA	AT-802	Minuteman
459	N70LA	AT-802	Aero Spray
473	N627LA	AT-802	New Frontier Av
474	N474H	AT-802	Henry's Flying Svc
477	N41525	AT-802	Western Pilot
478	N8502H	AT-802	Western Pilot
480	N8515G	AT-802	Aerial Timber
481	N8522Q	AT-802	Aero Spray
482	N619LA	AT-802	Aero Tech
483	N4149P	AT-802	Aero Tech
484	N166LA	AT-802	Aero Tech
485	N802LA	AT-802	Aero Tech
486	N8505E	AT-802	Aero Tech
487	N9197F	AT-802	Aero Tech
488	N332AT	AT-802	Aero Tech
491	N8509T	AT-802	Western Pilot Svc
492	N91357	AT-802	Western Pilot Svc
493	N428LA	AT-802	Western Pilot Svc
498	N398LA	AT-802AH	Western Pilot Svc
499	N870LA	AT-802AH	Western Pilot Svc
801	N325BM	AT-802	Fletcher FS
802	N802LA	AT-802	Aero Tech
803	N627LA	AT-802	New Frontier
807	N8523H	AT-802	New Frontier
810	N187LA	AT-802AH	Western Pilot Svc
819	N419FF	AT-802	Fletcher FS
820	N196LA	AT-802AH	Western Pilot Svc

Call Sign	Civil Registration	Aircraft	Comments
821	N207LA	AT-802AH	Western Pilot Svc
825	N802MA	AT-802	Minuteman
828	N428AE	AT-802	G.B. Aerial
830	N802SG	AT-802AH	Western Pilot Svc
839	N9190G	AT-802	Sly Av.
840	N427NS	AT-802	Queen Bee
841	N90802	AT-802	Queen Bee
845	N1531S	AT-802	Aerial Timber
846	N8515G	AT-802	Aerial TImber
847	N457AT	AT-802	Aerial Timber
848	N157LA	AT-802	Aerial Timber
849	N6162N	AT-802	Aerial Timber
858	N8522Q	AT-802	Aero Spray
859	N70LA	AT-802	Aero Spray
871	N802GB	AT-802F	G.B. Aerial
872	C-GLLG	AT-802F	New Frontier Av
873	N173TA	AT-802F	New Frontier Av
874	N174ML	AT-802F	New Frontier Av
880	N4169C	AT-802F	Queen Bee
882	N91092	AT-802F	Queen Bee
883	N9002K	AT-802F	Queen Bee
884	N9135F	AT-802F	Queen Bee
886	N1546H	AT-802F	Western Pilot Svc
888	N232AT	AT-802F	Aero Tech
890	N6159F	AT-802F	Aero Spray
891	N8512L	AT-802F	Aero Spray
892	N130BM	AT-802F	Evergreen
893	N802BM	AT-802F	Evergreen
894	N194BM	AT-802F	Evergreen
895	N195BM	AT-802F	Evergreen
898	N398LA	AT-802AH	Western Pilot Svc
899	N870LA	AT-802AH	Western Pilot Svc
Ayres Thrush			
411	N440AT	S2rT G-10	Western Pilot
412	N6125X	S2rT G-10	Western Pilot
413	N32973	S2rT T45	Western Pilot
414	N3299H	S2rT T45	Western Pilot
433	N4592T	S2rT T45	MJ Aviation
467	N660TR	S2rT 660	New Frontier Av
494	N22592	S2r G-10	Western Pilot Svc
Dromader			
455	N455TG	M18T	New Frontier Av
461	N4326M	M18T	New Frontier Av
463	N9152X	M18T	New Frontier Av
464	N92043	M18T	New Frontier Av
466	N466AT	M18T	New Frontier Av
469	N116BS	M18T	New Frontier Av
470	N13458	M18T	New Frontier Av
471	N7813T	M18T	New Frontier Av
475	N4QT	M18T	New Frontier Av
489	N110AF	M18T	New Frontier Av
Stearman PT-17/N2S			
E1	N75081	PT-17	
2 (?)	N63569	PT-17 (?)	

U.S. COMPANIES ASSOCIATED WITH AIR TANKERS

10 Tanker Air Carrier, Victorville, CA
Abe's Aerial Service, Safford, AZ
Aerial Applicators, Salt Lake City, UT
Aero Ag Inc., Medford, OR
Aero Atlas, Red Bluff, CA
Aero Enterprises, Troutdale, OR
Aero Enterprises Inc., Willows, CA
Aero-Flite/Aeroflite Inc., Cody, WY
Aero Union Corp., Chico, CA
Aero-Spray Inc., Yuma, CO
Aircraft Specialties, Mesa, AZ
Air Tractor Inc., Olney, TX
AJ Air Tankers, Van Nuys, CA
ARDCO, Tucson, AZ
Avery Aviation, Greybull, WY
Black Hills Aviation, Spearfish, SD, later Alamogordo, NM
Butler Aircraft Inc., Redmond, OR (Butler Farm Air)
Bruce Kinney/Kinney Air Tankers, Richey, MT
Cal-Nat Airways Inc., Grass Valley, CA
Capitol Air Inc., Carson City, NV
Central Air Service, East Wenatchee, WA, Maricopa, AZ
Central Air Service, Rantoul, KS
Central Oregon Aerial Co., Bend, OR
Cisco Aircraft, Lancaster, CA
Columbia Flying Service, Columbia, CA
Custom Farm Service of Montana, Richey, MT
Daro Inc., San Clemente, CA
Dennis G. Smilanich, Boise, ID
Donaire Inc., Phoenix, AZ
Edgar L. Thorsrud, Missoula, MT
Evergreen Air, Missoula, MT
Evergreen Aviation, Marana, AZ
Ewing Aviation, Los Angeles, CA
Fastaway Air Service, Long Beach, CA
Fire Flyers Inc., Reno, NV
Flick Aviation, La Grande, OR
Flight Enterprises, Prescott, AZ
George C. Abell, Topanga, CA
Globe Air Inc., Mesa, AZ
Hawkins and Powers Aviation, Greybull, WY
Hemet Valley Flying Service, Hemet, CA
Hillcrest Aviation, Boise, ID
Hills Flying Service, Chester and Placerville, CA
Idaho Air Tankers Inc., Boise, ID
Intermountain Aviation, Marana and Phoenix, AZ
Johnson Flying Service, Missoula, MT
Kem Air Inc., Worland, WY
Klamath Aircraft Inc., Klamath Falls, OR
Kreitzberg Aviation, Salem, OR
Liston Aircraft, Klamath Falls, OR
Lynch Air Tankers/Lynch Flying Service, Billings, MT
Moseley Aviation, Deer Valley, AZ
Neptune Aviation, Missoula, MT
Ostaire Flying Service, Ukiah, CA
P&B Aviation, Red Bluff, CA
P. F. Flickinger, Bayard, NM
Queen Bee, Rigby, ID
Rector Air Service, Chino, CA
Reeder Flying Service, Twin Falls, ID
Riverside Aircraft, Riverside, CA
Roberts Aircraft, Boise, ID
Rogue Flying Services, Medford, OR
Rosenbalm Aviation, Medford, OR
Routh Aircraft, Ramona and Paso Robles, CA
Sierra Aviation, Porterville, CA
Sis-Q Flying Service, Montague and/or Santa Rosa, CA
SLAFCO, Inc., Moses Lake, WA
Sonora Aviation, Tucson, AZ, and Carson City, NV
T&G Aviation, Chandler Field, AZ
TBM, Inc., Bakersfield and Tulare, CA
WAIG Aircraft, Inc., Tucson, AZ
Wen Inc., Porterville, CA
Wenairco, Wenatchee, WA
William A. Dempsey, East Wenatchee, WA
Willows Flying Service, Willows, CA
Wilson Aviation Industries, Lewiston, ID

AIR TANKER SAFE DROP HEIGHTS

The U.S. Forest Service produced a table in March 2000 that published safe heights for fixed-wing air tankers to make retardant drops. (*Aviation Tech Tips, Safe Drop Height for Fixed-Wing Airtankers,* by Greg Lovellette, March 2000.) As explained in a preface to the table, "Low-altitude retardant drops from fixed-wing airtankers may injure firefighters. Injuries are caused by flying debris (such as treetops or rocks) or when firefighters are knocked down. If the retardant has lost all of its forward momentum and is falling vertically as a heavy rain, the danger to firefighters is minimal." Safe drop height is defined as the distance below the air tanker at which retardant begins to fall vertically. The table lists air tankers according to the manufacturer of the tanks and drop gates. A number of air tankers, especially those modified by Aero Union, have multiple flow settings that affect drop height; high and low settings are sampled in this appendix. Since some manufacturers are also operators who additionally sell modifications to other operators, it becomes clearer to tie drop heights to the specific modified aircraft. In the intervening years, some of these historic air tanker models have been withdrawn from service.

Air Tanker Type	*Tank Volume*	*Flow Setting*	*Safe Drop Height (feet)*	
			Full Load	*Half Load*
ARDCO Aero-Flite DC-4	2,000	High	319	235
ARDCO Aero-Flite DC-4	2,000	Low	231	191
Aero Union Corp. C-130	3,000	9		249
Aero Union Corp. C-130	3,000	1/2	130	130
Aero Union Corp. P-3	3,000	Max	310	244
Aero Union Corp. P-3	3,000	1/2	140	140
Aero Union Corp. SP-2H	2,000	ED	234	173
Aero Union Corp. SP-2H	2,000	1/2	128	128
Aero Union Conventional C-54	2,000	High	260	207
Aero Union Conventional C-54	2,000	Low	202	180
Aero Union Conventional DC-7	3,000	High	284	221
Aero Union Conventional DC-7	3,000	Low	219	188
Hawkins and Powers C-130	3,000	High	287	219
Hawkins and Powers C-130	3,000	Low	241	196
Hawkins and Powers KC-97	3,200	High	323	237
Hawkins and Powers KC-97	3,200	Low	290	219
Hawkins and Powers KC-97	4,000	High	348	257
Hawkins and Powers KC-97	4,000	Low	298	228
Hawkins and Powers SPB4Y-2	2,200	High	265	208
Hawkins and Powers SPB4Y-2	2,200	Low	235	193
Hawkins and Powers SPB4Y-2	2,000	High	254	203
Hawkins and Powers SPB4Y-2	2,000	Low	223	187
Marsh Turbo Thrush 60/40 Door	380	40% Door	80	
Marsh Turbo Thrush 60/40 Door	380	60% Door	90	
Marsh Turbo Thrush 60/40 Door	380	Both Doors	100	
Minden Air Inc. P2V-7	2,450	High	288	221
Minden Air Inc. P2V-7	2,450	Low	231	170
Minden Air Inc. P2V-7	2,000	High	283	217
Minden Air Inc. P2V-7	2,000	Low	228	162
Neptune Inc. P2V-7 H&P P2V-7	2,450	High	288	219
Neptune Inc. P2V-7 H&P P2V-7	2,450	Low	215	183
Neptune Inc. P2V-5	2,450	High	291	221
Neptune Inc. P2V-5	2,450	Low	214	182
Pierce Turbo Thrush	450		80	
Snow Airtractor 802	800	Max	186	186
Snow Airtractor 802	800	1/2	92	92
Western Pilot Svcs Dromader Melex gate	500		140	
Western Pilot Svcs Dromader Transland	500		100	

AIR TANKER COMPARATIVE RATINGS

A 1959 USFS southern California study rated available air tanker aircraft on 11 criteria. The table of findings is excerpted here. The table should not be taken as defining the worthiness of a particular aircraft, but should be seen as a passing evaluation that, while interesting as a historical record, may not be universal in its conclusions.

Aircraft Type	*Min. Field Feet*	*Airspeed MPH*	*Capacity Gallons*	*Slow Speed Performance*	*Maneuverability*	*Visibility For Drop*	*Field Maintenance*	*Cost of Tanker*	*Pilot Skill*
N3N	1500	80	125	E	E	G	E	L	A
Twin Beech	4000	150	300	F	G	P	G	L	A
TBM	3000	170	600	E	E	G	P	L	A
F7F	3000	225	750	G	F-F*	E	F-P	H	S
PBY	3500	120	1000	E	F	F-P	E	H	A
B-25	5000	180	1000	P	P	F	P	L	S
B-26 (A-26)	5000	225	1200	P	P	F	G-P	L	S
PV-2	4000	175	1000	G	F	G	F	L	S
B-17	4000	170	1800	G	F-P	F-P	G	L	S
PB4Y-2	5000	175	2000	G	F	P	G	L	S
YC-97	5000	180	4500	G	P	E	P	H	S

E=Excellent G=Good F=Fair P=Poor H=High L=Low
Pilot Skill: A=Average Pilot Skill Required *Possibly a typo in the chart for F-P S=Special Pilot Skill Required

Appendix E

USFS REGIONS AND LETTER DESIGNATIONS

Tankers originally were assigned to a specific U.S. Forest Service Region, with a letter corresponding to each region. This gave rise to air tanker numbers with a prefix letter, as in E1, a PT-17 biplane in California's Region 5. Lists show some overlap of states in more than one region. USFS regions and their identifying letters were:

A = Region 1 (Montana, Idaho, Washington)
B = Region 2 (Colorado, Kansas, Nebraska, South Dakota, Wyoming)
C = Region 3 (Arizona, New Mexico)
D = Region 4 (Nevada, Utah, Idaho, Wyoming) — No letter assigned initially; by 1963, D was assigned
E = Region 5 (California)
F = Region 6 (Washington, Oregon)

GLOSSARY

Bird Dog: Common nickname for small spotter aircraft used to help direct air tankers on scene.

Borate Bomber: Catchy term coined by Los Angeles news media in the 1950s when borate was a promising new ingredient in retardant mix. Even though borate was quickly superseded by other chemicals, the term "borate bomber" remains in the lexicon.

DOI: Department of the Interior.

Mud: Slang term for fire retardant, probably originating with the use of bentonite clay in the 1950s. See also slurry.

SEAT: Acronym designating Single Engine Air Tanker, when single-engine aircraft were encouraged in the air tanker fleet in the 1990s.

Slurry: Term for fire retardant; more technically, a watery mix of largely undissolved chemical agent capable of retarding fire, as with bentonite slurry or borate slurry.

USFS: United States Forest Service.

Viscous: Fluid with consistency thicker than water.

VLAT: Very Large Air Tanker.

Wetting Agent: Chemical added to water to reduce surface tension to promote dispersal and penetration.

Wet Water: Water treated with a wetting agent; useful in some firefighting applications.

ENDNOTES

1 "What's News Across the Nation," *American Forests,* American Forestry Association, Vol. 60, No. 3, March 1954.
2 "Forest Fire Air Attack System," William T. Larkins, American Aviation Historical Society *Journal,* Vol. 9, No. 3, Fall 1964.
3 "Army Aircraft to Fight Forest Fires," *American Forestry,* American Forestry Association, Vol. 25, No. 305, May 1919.
4 *Ibid.*
5 "Air Pilots to Spot Forest Fires," *American Forestry,* American Forestry Association, Vol. 26, No. 317, May 1920.
6 "California Airplane Fire Patrol Succeeds," *American Forestry,* American Forestry Association, Vol. 27, No. 325, January 1921.
7 *Ibid.*
8 "The Forest Fire War in California," George C. Henderson, *American Forests,* circa 1924.
9 *Ibid.*
10 "Notes," *Journal of Forestry,* Society of American Foresters, Vol. XX, 1922.
11 "The Beginning of the Washington Air National Guard," John E. Dean, *Pacific Northwest Aviation Historical Foundation Journal,* Vol. 4, 1973.
12 Letter, John R. White, Superintendent, Giant Forest, to Director, National Park Service, Washington, D.C., 10 September 1926.
13 *Aircraft Year Book—1928,* Aeronautical Chamber of Commerce of America, Inc., New York City, New York, 1928.
14 *Ibid.*
15 "Forest Fire Air Attack System," William T. Larkins, American Aviation Historical Society *Journal,* Vol. 9, No. 3, Fall 1964.
16 "Rain From Heaven—At Will?," *USFS Northern Region News,* Vol. VII, No. 6, 21 March 1935, Missoula, Montana.
17 "Forest Fire Air Attack System," William T. Larkins, American Aviation Historical Society *Journal,* Vol. 9, No. 3, Fall 1964.
18 "Re-thinking Forest Fire Control," by Roy Headley, Northern Rocky Mountain Forest and Range Experimental Station, 15 March 1943.
19 *Ibid.*
20 *Ibid.*
21 *Ibid.*
22 "Forest Fire Air Attack System," William T. Larkins, American Aviation Historical Society *Journal,* Vol. 9, No. 3, Fall 1964.
23 *Ibid.*
24 *Ibid.*
25 *Report of Aerial Bombing Evaluation Board,* Missoula, Montana, 22 August 1947.
26 *Ibid.*
27 "Forest Fire-Fighting Retardants in the United States," Merle S. Lowden, *Canadian Pulp and Paper Association* presentation paper, 20 March 1962.
28 "Aerial Firefighting—Bulk Water Drop by Fixed Wing Aircraft," *Firestop,* Progress Report No. 9, 15 June 1955.
29 *Ibid.*

30 "Forest Fire Air Attack System," William T. Larkins, American Aviation Historical Society *Journal,* Vol. 9, No. 3, Fall 1964.
31 "Tank Installation in T.B.M. Air Tanker," *Technical Equipment Report No. 5100-11,* Arcadia Development Center, USFS, May 1960.
32 "Forest Service—Department of Agriculture," *Aircraft Year Book,* 1957-58, American Aviation Publications, Inc., Washington, D.C.
33 "Fewer Pest Control Programs Mean Tighter Competition for Sprayers," *Aviation Week Including Space Technology,* 23 March 1959.
34 "Evaluation of a Grumman TBM-3 Air Tanker," ATO TR-62-11, United States Army Aviation Test Office, Edwards AFB, California, November 1962.
35 Advertisement, "Firebrake ® lets you 'buy time' at bargain rates," in *American Forests* magazine, August 1960.
36 "Forest Fire-Fighting Retardants in the United States," Merle S. Lowden, *Canadian Pulp and Paper Association* presentation paper, 20 March 1962.
37 *Chemicals for Forest Fire Fighting,* National Fire Protection Association International, 1963.
38 "Uncle Sam's Flying Fire Department," by Keith Monroe, *American Forests,* American Forestry Association, Vol. 67, No. 9, September 1961.
39 *Air Drop Tests—Willows, Santa Ana, Ramona 1955-59,* A Report For The California Air Attack Coordinating Committee, by James Davis.
40 "A New Forest Fire Retardant—Swelling Bentonite Clay," by Clinton B. Phillips and Harrry R. Miller, Jr., *Journal of Forestry,* Society of American Foresters, Vol. 58, No. 12, December 1960.
41 *Ibid.*
42 *Air Drop Tests—Willows, Santa Ana, Ramona 1955-59,* A Report For The California Air Attack Coordinating Committee, by James Davis.
43"Forest Fire Air Attack System," William T. Larkins, American Aviation Historical Society *Journal,* Vol. 9, No. 3, Fall 1964.
44 "Put Out That Fire With Seaweed!" *Fire Control Experiments,* No. 1, February 1961, California Division of Forestry.
45 "Forest Fire-Fighting Retardants in the United States," Merle S. Lowden, *Canadian Pulp and Paper Association* presentation paper, 20 March 1962.
46 Advertisement, "The best way to tame a wildfire is with chemicals," Monsanto, circa 1977.
47 Information about Fire-Trol was posted in Frequently Asked Questions (FAQs), Fire-Trol Web site, www.firetrolholdings.com as of this writing.
48 *Ibid.*
49 Information about Firefoam was posted in Products, Fire-Trol Web site, www.firetrolholdings.com as of this writing.
50 *Chemicals for Forest Fire Fighting,* National Fire Protection Association International, October 1964 Supplement.
51 *Wildland Fire Suppression Tactics Reference Guide,* National Wildfire Coordinating Group, April 1996.
52 *Test Report—High Altitude Drop Demonstration Ryan Field, Hemet, California, 12 November 1971,* prepared by E. B. Anderson, Aerospace Report No. ATR-72(S7264)-3.
53 *Ibid.*
54 *Ibid.*
55 *Three New Air Tankers Cascade Water Over Orange County Airport,* Report by Arcadia Development Center, USFS, 1957.
56 *Hazards to Ground Personnel from Air Drops of Fire Retardants,* USFS Arcadia Equipment Development Center, Arcadia, California, June 1961.
57 *Ibid.*
58 *Ibid.*
59 *Corrosion of Air Tankers by Fire Retardants,* by James B. Davis and Clinton B. Phillips, A report for the California Air Attack Coordinating Committee—1965, printed and distributed by the California Division of Forestry.
60 *Ibid.*
61 *Ibid.*
62 *Air Attack—Southern California, 1959,* U.S. Department of Agriculture—Forest Service—California Region.
63 *Ibid.*
64 *Ibid.*
65 *Ibid.*
66 "Air Tankers in Southern California Fires… effectiveness in delivering retardants rated," by Theodore G. Storey and Leon W. Cooley, Jr., *Pacific Southwest Forest & Range Experiment Station, 1967.*
67 *Ibid.*
68 *Ibid.*
69 "Aerial Firefighting—Bulk Water Drop by Fixed Wing Aircraft," *Firestop,* Progress Report No. 9, 15 June 1955.
70 *Ibid.*
71 "Forest Fire Air Attack System," William T. Larkins, American Aviation Historical Society *Journal,* Vol. 9, No. 3, Fall 1964.
72 *Ibid,* and "Uncle Sam's Flying Fire Department," by Keith Monroe, *American Forests,* American Forestry Association, Vol. 67, No. 9, September 1961, and *CDF Aviation Management History* (undated) by California Department of Forestry and Fire Protection.
73 "Evaluation of a Grumman TBM-3 Air Tanker," ATO TR-62-11, United States Army Aviation Test Office, Edwards AFB, California, November 1962.
74 *Ibid.*
75 *Ibid.*
76 *Ibid.*
77 *Ibid.*
78 *Ibid.*
79 *Ibid.*
80 *Ibid.*
81 *Ibid.*
82 *Testing the TBM Aerial Tanker in the Southeast,* by Theodore G. Storey, George W. Wendel, and Anthony T. Altobellis, Station Paper No. 101, U.S. Department of Agriculture—Forest Service, Southeastern Forest Experiment Station, Asheville, North Carolina, July 1959.
83 *Ibid.*
84 "Forest Fire Air Attack System," William T. Larkins, American Aviation Historical Society *Journal,* Vol. 9, No. 3, Fall 1964.
85 "TB-25N Airtanker Flight Evaluation," Special Report Under ED&T-1114, by Arcadia Equipment Development Center and U.S. Army Aviation Test Office, TMC/Edwards AFB; U.S. Department of Agriculture/Forest Service/Washington, D.C., February 1963.
86 *Ibid.*

87 *Ibid.*
88 *Ibid.*
89 *Ibid.*
90 *Ibid.*
91 *Ibid.*
92 *Ibid.*
93 *Final Cut—The Post-War B-17 Flying Fortress: The Survivors,* Scott A. Thompson, Pictorial Histories Publishing Company, Missoula, Montana, Fourth Printing December 1995.
94 *Ibid.*
95 *Ibid,* and *Warbirds Worldwide Directory,* John Chapman and Geoff Goodall, Warbirds Worldwide Limited, Mansfield, England, 1989.
96 National Transportation Safety Board report, DEN71FXA05, Aircraft: Boeing TB-17F, registration: N1340N.
97 National Transportation Safety Board report, SEA71DVQ03, Aircraft: Convair PB4Y-2, registration: N3191G.
98 "Safety Concerns Ground Aerial Firefighting Tankers," by William B. Scott, *Aviation Week and Space Technology,* 16 December 2002.
99 *Forest Firefighting External Tank for DeHavilland Beaver DHC-2 Seaplane,* Technical Equipment Report No. 5700-4, Ely Service Center, Forest Service, U.S. Department of Agriculture, 15 February 1962.
100 *Ibid.*
101 *Ibid.*
102 *The DC-3—50 Years of Legendary Flight,* Peter M. Bowers, Aero/Tab Books, Blue Ridge Summit, Pennsylvania, 1986.
103 *Interagency Airtanker Base Operations Guide, Appendix G: Airtanker Identification,* National Interagency Aviation Council, NFES 2271, May 2007.
104 *Interagency Airtanker Base Operations Guide, Appendix G: Airtanker Identification,* National Interagency Aviation Council, NFES 2271, May 2007.
105 *Federal Aerial Firefighting: Assessing Safety and Effectiveness,* Blue Ribbon Panel Report to the Chief, USDA Forest Service and Director, USDI Bureau of Land Management, December 2002.
106 *Interagency Airtanker Base Operations Guide, Appendix G: Airtanker Identification,* National Interagency Aviation Council, NFES 2271, May 2007.
107 *Douglas DC-6 & DC-7 Retardant Tank System Operations Manual,* Aero Union Corporation, Chico, California. (Undated).
108 *Ibid.*
109 *Ibid.*
110 *Fairchild C-82 Packet and C-119 Flying Boxcar,* Alwyn T. Lloyd, Midland Publishing, (Aerofax, Ian Allan), Hinckley, England, 2005.
111 "Forest Fire Air Attack System," William T. Larkins, American Aviation Historical Society *Journal,* Vol. 9, No. 3, Fall 1964.
112 *North American AJ-1 Savage, Naval Fighters Number Twenty Two,* Steve Ginter, Simi Valley, California, 1992.
113 *Ibid.*
114 *Air Tanker Helicopter & Large Transport Information—National and Regional,* USFS-BIFC, Boise, Idaho, June 1983.
115 *Air Tanker—Evaluation, P2V-5F—N126Z.* This is a series of mostly unsigned papers and technical data generated after tests of the first P2V air tanker in 1969, and furnished from Boise Interagency Fire Center historical documents.
116 *Ibid.*
117 *Ibid.*
118 *Flight Test Report,* P2V-5F Airplane Number 126Z, Company Pilot—Bill Rosenbalm, 15 October 1969, bundled as part of *Air Tanker—Evaluation, P2V-5F—N126Z.* This is a series of mostly unsigned papers and technical data generated after tests of the first P2V air tanker in 1969, and furnished from Boise Interagency Fire Center historical documents.
119 *Air Tanker—Evaluation, P2V-5F—N126Z.* This is a series of mostly unsigned papers and technical data generated after tests of the first P2V air tanker in 1969, and furnished from Boise Interagency Fire Center historical documents.
120 *Ibid.*
121 *Ibid.*
122 *Ibid.*
123 Letter, by Lou Remschner, 7 January 1971, bundled as part of *Air Tanker—Evaluation, P2V-5F—N126Z.* This is a series of mostly unsigned papers and technical data generated after tests of the first P2V air tanker in 1969, and furnished from Boise Interagency Fire Center historical documents.
124 Letter, by Richard W. Bradley, Lt. Col., USAF (Retired), to Rosenbalm Aviation, 7 January 1971, bundled as part of *Air Tanker—Evaluation, P2V-5F—N126Z.* This is a series of mostly unsigned papers and technical data generated after tests of the first P2V air tanker in 1969, and furnished from Boise Interagency Fire Center historical documents.
125 Letter, "Restricted Operating Limitations—Multiengine Aircraft; Aircraft: Lockheed; Model: SP-2H (P2V-7); Serial: 8013; Registration: N14448, 13 June 1973, From: Federal Aviation Administration General Aviation District Office, Helena, Montana, by Richard F. Brodowy, General Aviation Maintenance Inspector.
126 *Lockheed Model SP-2H (P2V-7) Air Tanker Conversion—Johnson Flying Service, Inc.—Flight Manual Supplement #1,* 11 July 1972.
127 "Ground Pattern Performance of the Neptune P2V-7 Airtanker," *Airtanker Drop Guide,* by Paul Solarz, Program Leader, and Cammie Jordan, Project Assistant, 0057-2848-MTDC, September 2000.
128 *Ibid.*
129 *Ibid.*
130 "Ground Pattern Performance of the Airspray Electra L-188 with Aero Union Constant Flow Tank," *Airtanker Drop Guides,* by Paul Solarz, Program Leader, and Cammie Jordan, Project Assistant, 0057-2851-MTDC, September 2000.
131 *Ibid.*
132 *Interagency Airtanker Base Operations Guide, Appendix G: Airtanker Identification,* National Interagency Aviation Council, NFES 2271, May 2007.
133 Letter, Maj. Carol A. Habgood, USAF Office of Information to the author, 8 August 1977, subject: Air National Guard/Air Force Reserve C-130 firefighting operations.

134 *Report on Tests of the FMC Corporation's Modular Airborne Fire Fighting System—(MAFFS),* by Charles W. George, October 1971.

135 "Air Force Hercules Freighters Proposed as Emergency Forest Fire Fighter/Tankers," Lockheed news release 30 September 1971 Burbank, California.

136 *Lockheed C-130 Hercules and Its Variants,* by Chris Reed, Schiffer Military History, Atglen, Pennsylvania, 1999.

137 "U.S. Forest Service accepts first C-130 MAFFS II/AFFS Airborne Fire Fighting System," news release, Aero Union Corp., 27 July 2007.

138 *Ibid.*

139 "Forest Fire Air Attack System," by William T. Larkins, American Aviation Historical Society *Journal*, Fall 1964.

140 *Interagency Airtanker Base Operations Guide, Appendix G: Airtanker Identification,* National Interagency Aviation Council, NFES 2271, May 2007.

141 *CDF Aviation Management History* (undated) by California Department of Forestry and Fire Protection.

142 Bulletin, 19 July 2002, by Aerial Firefighting Industry Association (AFIA).

143 Bulletin, 4 December 2002, by Aerial Firefighting Industry Association (AFIA).

144 *Federal Aerial Firefighting: Assessing Safety and Effectiveness,* Blue-Ribbon Panel Report to the Chief, USDA Forest Service and Director, USDI Bureau of Land Management, December 2002.

145 *Ibid.*

146 *Ibid.*

147 *Ibid.*

148 *Ibid.*

149 *Ibid.*

150 *Ibid.*

151 *Ibid.*

152 *Ibid.*

153 *Ibid.*

154 *Ibid.*

155 *Ibid.*

156 *Ibid.*

157 "U.S.D.A. Forest Service and Department of the Interior agencies cancel large airtanker contract," National Interagency Fire Center news release, 10 May 2004.

158 News Release, "Sandia helps return firefighting air tankers to service with improved inspection programs," Sandia National Laboratories, 27 March 2003.

159 From interview notes made by the author with Asher Williams, USFS National Aviation Operations Officer, 13 November 2006.

160 *Ibid.*

161*Addendum to Federal Fire & Aviation Operations 2004 Action Plan,* USDA Forest Service/Department of the Interior, Fire and Aviation Management, 25 May 2004.

162 *Ibid.*

163 *Ibid.*

164 *Limited Number of Large Airtankers Recalled to Service,* NIFC News, National Interagency Fire Center, Boise, Idaho, 2 July 2004.

165 *Federal Agencies' Aerial Resources are Ready for 2005 Fire Season,* Department of the Interior press release, 26 May 2005.

166 *USFS Very Large Aerial Tanker Operational Test and Evaluation,* Summary Report, 2 March 2009.

167 *Ibid.*

168 *Ibid.*

169 *Ibid.*

170 *Ibid.*

171 Evergreen International Aviation Web site at www.evergreenaviation.com

172 News release, "Evergreen Supertanker Ready to Fight Wildfires—The B747 Supertanker is certified to fly by Interagency Air Tanker Board," Evergreen International Aviation, 24 March 2009.

173 *Ibid.*

174 Interview, author with Penn Stohr, 23 October 2009.

175 Fact sheet, *DC-10 Airtanker—Tanker 910,* California Department of Forestry & Fire Protection, www.fire.ca.gov, June 2007.

176 *Ibid.*

177 *Ibid.*

178 "AT-802F Single Engine Air Tanker," Web information, www.airtractor.com circa 1 May 2009.

179 "Fire Boss LLC Announces America's First 'Seat' Rated Fire Boss Pilots," 9 November 2006, www.firebossllc.com circa 1 May 2009.

180 See also various source materials available at www.firebossllc.com circa 1 May 2009.

181 "AT-802F 'Fire Boss' SEATs Sold to Croatia," Web news release, www.airtractor.com circa 1 May 2009.

182 "Wipaire 'Fire Boss' Federally Carded 4-19-08, Web news release, www.firebossllc.com circa 1 May 2009.

183 *Jane's All The World's Aircraft, 1995-96,* Jane's Publishing Co. Ltd., Coulsdon, Surrey, UK, 1995.

184 *Jane's All The World's Aircraft, 1986-87,* Jane's Publishing Co. Ltd., London, England, 1986.

185 "Ground Pattern Performance of the Western Pilot Services Dromader," *Airtanker Drop Guides,* USFS Technology and Development Program, 0057-2834-MTDC.

186 *The Dromader M-18 Aircraft: Its Role as an Air Tanker,* by E. Stechisen and G. Halicki, Information Report PI-X-40, Petawawa National Forestry Institute, Canadian Forestry Service, 1984.

187 *Jane's All The World's Aircraft, 1986-87,* Jane's Publishing Co. Ltd., London, England, 1986, and *Jane's All The World's Aircraft, 1995-96,* Jane's Publishing Co. Ltd., Coulsdon, Surrey, UK, 1995.

188 "Oklahoma Uses Airplanes to Fight Range Fires," (brochure), Oklahoma State University Leaflet L-108, 1964.

189 "Interagency Airtanker Board Charter," National Interagency Aviation Council, Internet information circa April 2009.

190 Interagency Airtanker Board Briefing Paper, Topic: Interagency Airtanker Board (IAB); Issue: Understanding the role and responsibility of the IAB, 6 June 2006.

191 *Ibid.*

192 *Ibid.*

193 *Ibid.*

194 *Ibid.*

195 *Ibid.*

196 *Ibid.*

197 *Ibid.*

198 *Ibid.*

199 *Ibid.*

200 *Ibid.*

201 "Argentinians Adapt Arava into Water Bomber." *Aviation Week and Space Technology,* 20 October 1980.
202 From interview notes made by the author with Asher Williams, USFS National Aviation Operations Officer, 13 November 2006.
203 "Air Force Hercules Freighters Proposed as Emergency Forest Fire Fighters/Tankers," Lockheed news release 30 September 1971 Burbank, California.
204 "Forest Fire Air Attack System," by William T. Larkins, American Aviation Historical Society *Journal*, Fall 1964.
205 *Darkly Dangerous—The Northrop P-61 Black Widow Night Fighter,* by Frederick A. Johnsen, Bomber Books, Tacoma, Washington, 1981, and "Forest Fire Air Attack System," by William T. Larkins, American Aviation Historical Society *Journal*, Fall 1964.
206 "Forest Fire Air Attack System," by William T. Larkins, American Aviation Historical Society *Journal*, Fall 1964.
207 *Darkly Dangerous—The Northrop P-61 Black Widow Night Fighter,* by Frederick A. Johnsen, Bomber Books, Tacoma, Washington, 1981. Names of fallen aerial firefighters have been memorialized by the Associated Airtanker Pilots at www.airtanker.org as of 2009.
208 Information on the Associated Airtanker Pilots Memorial Fund was posted online at www.airtanker.org as of 2009.
209 "The Use of Seaplanes in Forest Mapping," by Ellwood Wilson, *Journal of Forestry,* Society of American Foresters, Vol. XVIII, No. 1, January 1920.
210 "Forestry From the Air," by Stanley Frost, *American Forestry,* American Forestry Association, Vol. 27, No. 329, May 1921.
211 *Ibid.*
212 "The Use of Seaplanes in Forest Mapping," by Ellwood Wilson, *Journal of Forestry,* Society of American Foresters, Vol. XVIII, No. 1, January 1920.
213 "Ontario's Aerial Fire Fighters," by James Montagnes, *American Forests,* American Forestry Association, Vol. 69, No. 5, May 1963.
214 From miscellaneous clippings in USFS historical files at Missoula, Montana.
215 "Forest Fire Air Attack System," by William T. Larkins, American Aviation Historical Society *Journal*, Fall 1964.
216 *Ibid.*
217 "Aerial 'Water Bombing,'" *Firemen,* March 1954.
218 "Airborne Bucket Brigade," *Aircraft—Canada's Aero Trade Magazine,* October 1957.
219 *Ibid*, and "Forest Fire Air Attack System," by William T. Larkins, American Aviation Historical Society *Journal*, Fall 1964.
220 "Ontario's Aerial Fire Fighters," by James Montagnes, *American Forests,* American Forestry Association, Vol. 69, No. 5, May 1963.
221 *Development of Aircraft Bombing Systems,* J. K. Hawkshaw, Chief Aeronautical Engineer, Field Aviation Company Limited, February 1969.
222 "PBY Water Bombers Battle Forest Fires," *Aviation Week and Space Technology,* 3 August 1964.
223 *Ibid.*
224 *Ibid.*
225 *Fire Control Manual,* Forest Industries Flying Tankers Limited, Sproat Lake, Port Alberni, B.C., May 1966.
226 *Ibid.*
227 "Canadian Teamwork in Forest Firebombing," *Canadian Wings.*
228 *Canadian Aircraft Since 1909,* K. M. Molson and H. A. Taylor, Putnam, London, 1982.
229 *Ibid.*
230 *Jane's All the World's Aircraft, 1990-91,* Mark Lambert, Editor-In-Chief, Jane's Defence Data, Surrey, UK, 1990.
231 *Ibid.*
232 "Forest Fire Fighting Services Showing Renewed Interest in Water-Based Aircraft," *Canadian Aircraft Operator,* 16 May 1977.
233 *Ibid.*
234 *Ibid.*
235 *Ibid.*
236 *Ibid.*
237*Jane's All the World's Aircraft, 1989-90,* John W. R. Taylor, Editor-In-Chief, Jane's Defence Data, Surrey, UK, 1989.
238 "CV580 Airtanker" web information, Kelowna Flightcraft and Conair Aviation, available as of 26 April 2009.
239 *Interagency Airtanker Base Operations Guide,* NFES 2271, Sponsored by National Interagency Fire and Aviation Executive Board, May 2007.
240 *Interagency Airtanker Base Operations Guide,* NFES 2271, Sponsored by National Interagency Fire and Aviation Executive Board, May 2007.
241 *Ibid.*
242 *Ibid.*
243 *Ibid.*
244 *Ibid.*
245 *Ibid.*
246 *Ibid.*
247 *Ibid.*
248 From a letter written by Joe Ely, U.S. Forest Service, in 1981 and preserved by Irene and George Nolta.
249 *Wildland Fire Suppression Tactics Reference Guide,* National Wildfire Coordinating Group, April 1996.
250 *Ibid.*
251 *Ibid.*
252 *Ibid.*
253 *Ibid.*
254 *Ibid.*
255 "Uncle Sam's Flying Fire Department," by Keith Monroe, *American Forests,* American Forestry Association, Vol. 67, No. 9, September 1961.
256 *CDF Aviation Management History* (undated) by California Department of Forestry and Fire Protection.
257 *Federal Aerial Firefighting: Assessing Safety and Effectiveness,* Blue-Ribbon Panel Report to the Chief, USDA Forest Service and Director, USDI Bureau of Land Management, December 2002.
258 "Chico School Teaches Firefighting From Air," by Lynda Kenyon, *Sacramento Bee,* 25 July 1976.
259 Interview, author with Penn Stohr, 23 October 2009.
260 *Ibid.*
261 *Ibid.*
262 *Ibid.*
263 *Ibid.*
264 *Ibid.*
265 *Ibid.*
266 "The Proud Bird," by Cal Butler, *Ag-Pilot International,* September 1985.

INDEX

More Great Titles From Specialty Press

AMARG: America's Military Aircraft Boneyard *by Nick Veronico* This country's largest military aircraft storage center began as a small desert site holding bombers and transports in reserve for possible future use. It later became more of a salvage and parts recovery operation, and in many cases, a final resting place known as "the boneyard." In the 1950s and 1960s, with new wars erupting in Korea and Vietnam, certain aircraft stored in this desert center were once again in demand, and this famed storage and salvage facility in Tucson, Arizona, answered the call. Numerous photographs taken both from the air and on the ground show the reader vistas of the 4,000 airplanes stored at this site, while a detailed appendix gives a comprehensive listing of all the aircraft types currently at AMARG (Aerospace Maintenance and Regeneration Group). Softbound, 11 x 8-1/2 inches, 144 pages, 50 b/w & 350 color photos. ***Part #SP139***

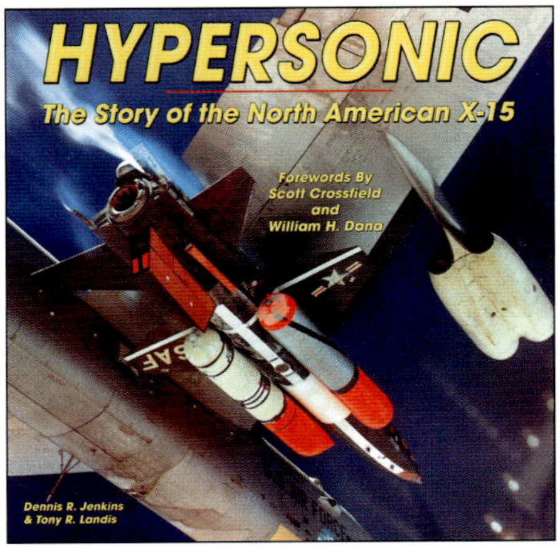

HYPERSONIC: The Story of the North American X-15 *by Dennis R. Jenkins & Tony R. Landis* This is the most extensively researched history of the X-15 program yet produced, written with the cooperation of surviving X-15 pilots as well as many other program principals. It covers the tasks of converting and testing the B-52 carrier airplanes, building the first full-pressure suits to protect the pilot, building the first engineering mission simulators, acquiring the remote lakebed landing sites, and building the radar range. It also covers the flight program in detail, including the most authoritative flight log ever assembled. Softbound, 9 x 9 inches, 276 pages, 500 b/w & 50 color photos, 100 drawings. ***Part #SP131***

LOCKHEED BLACKBIRD FAMILY: A-12, YF-12, D-21/M-21 & SR-71 Photo Scrapbook *Tony Landis* The Lockheed family of A-12, YF-12, D-21/M-21, and SR-71 Blackbirds are to this day the highest-performance jet-powered airplanes ever flown. Although no longer operational, A-12s and SR-71s flew for nearly three decades at speeds in excess of Mach 3 and altitudes of up to 90,000 feet. This book is compiled of many never-before-published photos coupled with de-classified information recently released by the CIA. Softbound 8-1/2 x 11 inches, 128 pages, 425 illustrations. ***Part #SP151***

X-PLANE CRASHES: Exploring Experimental, Rocket Plane, and Spycraft Incidents, Accidents and Crash Sites *by Peter W. Merlin & Tony Moore* Known as "The X-Hunters," the authors have located more than 100 crash sites of exotic aircraft from Edwards Air Force Base and Area 51. Together, they have recovered parts of supersonic rocket planes, stealthy spy craft, and vehicles that have reached the edge of space. Each story in the book profiles an unusual aircraft and the brave men who flew them. The authors examine the contributing causes of each crash and use then-and-now photographs to illustrate their findings. Aircraft profiled include the YB-49 and a pair of N9M flying wings, X-1A, X-1D, VB-51, XB-70, SR-71, YF-12, U-2 prototype, and many more. Hardbound, 8-1/2 x 11 inches, 160 pages, 50 color & 225 b/w photos. ***Part #SP121***

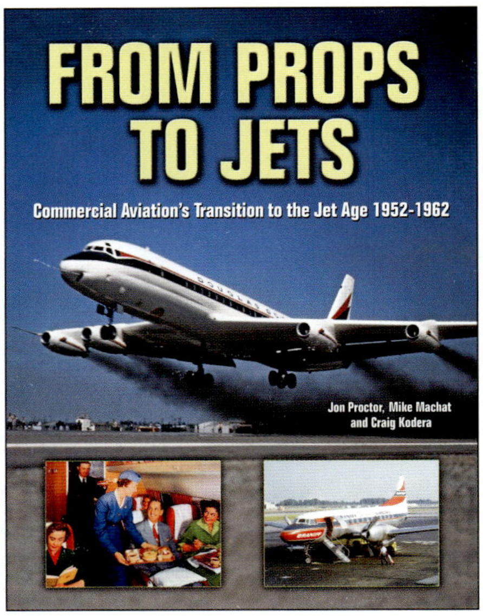

FROM PROPS TO JETS: Commercial Aviation's Transition to the Jet Age 1952–1962 *Jon Proctor, Mike Machat, & Craig Kodera* Although the years from 1952 to 1962 saw the rise and dominance of luxurious piston-powered airliners like the Douglas DC-7, Lockheed Constellation, and Boeing Stratocruiser, revolutionary new jet airliners soon rendered them obsolete. Aircraft like the Boeing 707, Douglas DC-8, and Convair 880 sliced travel times around the world in half as the traveling public enthusiastically embraced the new and very futuristic jet age. This book takes an in-depth look at all the aircraft that made commercial jet flight possible, comparing these radical new jetliners with the classic propeller-driven aircraft that preceded them. Hardbound, 8-1/2 x 11 inches, 160 pages, 120 bw & 60 color photos, 20 illustrations. ***Part #SP146***

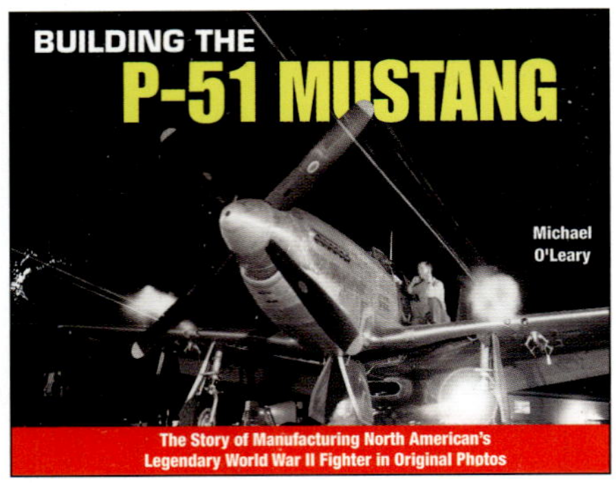

BUILDING THE P-51 MUSTANG IN ORIGINAL FACTORY PHOTOS: The Manufacturing Story of North American's Legendary World War II Fighter *Michael O'Leary* Now, for the first time, here is a major work devoted strictly to telling the story of how this legendary airplane was designed and built using rare original factory photographs, documents, and unique engineering information. The author uses more than 300 original photos culled from his personal archive of official North American and USAAF images, many of which have never before been seen in any publication whatsoever. This book will provide a vital "missing link" in the saga of this famed World War II aircraft. Hardbound, 8-1/2 x 11 inches, 240 pages, 300 b/w photos & 50 b/w illustrations. ***Part #SP152***

Specialty Press
39966 Grand Avenue, North Branch, MN 55056
Telephone: (651) 277-1400 or (800) 895-4585 Fax: (651) 277-1203
www.specialtypress.com